AQA Science

Exclusively endorsed and approved by AQA

Revision Guide

John Scottow

Series Editor: Lawrie Ryan

GCSE Chemistry

Nelson Thornes

Published in 2006 by:
Nelson Thornes Ltd
Delta Place
27 Bath Road
CHELTENHAM
GL53 7TH
United Kingdom

08 09 10 / 10 9 8 7 6 5 4

A catalogue record for this book is available from the British Library

ISBN 978 0 7487 8314 4

Cover photographs: flames by Photodisc 29 (NT); chemical crystals by Photodisc 4 (NT); chemical flasks by Photodisc 54 (NT)
Cover bubble illustration by Andy Parker

Illustrations by Bede Illustration
Page make-up by Wearset Ltd

Printed and bound in Croatia by Zrinski

Acknowledgements

Alamy/Adam Woolfit 22b, /**Adrian Sherratt** 4m, /**Rob Walls** 20t; **Arm & Hammer** 85m; **Corel 250 (NT)** 23br; **Corel 357 (NT)** 71br, 86tr; **Corel 437 (NT)** 5l; **Corel 710 (NT)** 56br; **Digital Vision 1 (NT)** 59t; **Digital Vision 15 (NT)** 12m; **Martyn F. Chillmaid** 74br; **Perkin Elmer** 100bm; **Photodisc 71 (NT)** 91tr; **Royal Society of Chemistry** 73tl; **Science Photo Library/Andrew Lambert Photography** 19r, 65tl, 91br, /**Andrew McClenaghan** 24m, /**Astrid & Hanns-Frieder Michler** 84br, /**Cordelia Molloy** 17r, 57tl, /**GECO UK** 99bl, /**John McLean** 101tl, /**Lawrence Lawry** 20mr, /**Martin Bond** 13r, /**Pascal Goetgheluck** 8l, /**TEK Image** 100br; **Stockbyte 29 (NT)** 20bl

Many thanks for the contributions made by Paul Lister and Patrick Fullick.

Picture research by Stuart Silvermore, Science Photo Library and johnbailey@ntlworld.com.

Every effort has been made to trace all the copyright holders, but if any have been overlooked the publisher will be pleased to make the necessary arrangements at the first opportunity.

How to answer questions

Question speak

Command word or phrase	What am I being asked to do?
compare	State the similarities and the differences between two or more things.
complete	Write words or numbers in the gaps provided.
describe	Use words and/or diagrams to say how something looks or how something happens.
describe, as fully as you can	There will be more than one mark for the question so make sure you write the answer in detail.
draw	Make a drawing to show the important features of something.
draw a bar chart / graph	Use given data to draw a bar chart or plot a graph. For a graph, draw a line of best fit.
explain	Apply reasoning to account for the way something is or why something has happened. It is not enough to list reasons without discussing their relevance.
give / name / state	This only needs a short answer without explanation.
list	Write the information asked for in the form of a list.
predict	Say what you think will happen based on your knowledge and using information you may be given.
sketch	A sketch requires less detail than a drawing but should be clear and concise. A sketch graph does not have to be drawn to scale but it should be the appropriate shape and have labelled axes.
suggest	There may be a variety of acceptable answers rather than one single answer. You may need to apply scientific knowledge and/or principles in an unfamiliar context.
use the information	Your answer **must** be based on information given in some form within the question.
what is meant by	You need to give a definition. You may also need to add some relevant comments.

Diagrams

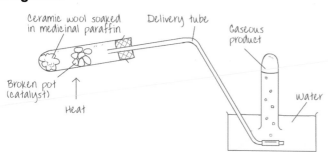

Things to remember:

- Draw diagrams in pencil.
- The diagram needs to be large enough to see any important details.
- Light colouring could be used to improve clarity.
- The diagram should be fully labelled.
- Label lines should be thin and end at the point on the diagram that corresponds to the label.

How long should my answer be?

Things to consider:

1 How many lines have been given for the answer?
   - One line suggests a single word or sentence. Several lines suggest a longer and more detailed answer is needed.

2 How many marks is the answer worth?
   - There is usually one mark for each valid point. So for example, to get all of the marks available for a three mark question you will have to make three different, valid points.

3 As well as lines, is there also a blank space?
   - Does the question require you to draw a diagram as part of your answer?
   - You may have the option to draw a diagram as part of your answer.

Graphs

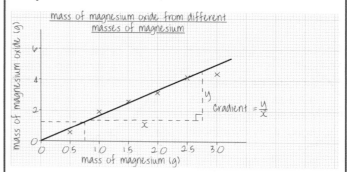

Things to remember:

- Choose sensible scales so the graph takes up most of the grid.
- Don't choose scales that will leave small squares equal to 3 as it is difficult to plot values with sufficient accuracy on such scales.
- Label the axes including units.
- Plot all points accurately by drawing small crosses using a fine pencil.
- Don't try to draw a line through every point. Draw a line of best fit.
- A line of best fit does not have to go through the origin.
- When drawing a line of best fit, don't include any points which obviously don't fit the pattern.
- The graph should have a title stating what it is.
- To find a corresponding value on the y-axis, draw a vertical line from the x-axis to the line on the graph, and a horizontal line across to the y-axis. Find a corresponding value on the x-axis in a similar way.
- The gradient or slope of a line on a graph is the amount it changes on the y-axis divided by the amount it changes on the x-axis. (See the graph above.)

Calculations

- Write down the equation you are going to use, if it is not already given.
- If you need to, rearrange the equation.
- Make sure that the quantities you put into the equation are in the right units. For example you may need to change centimetres to metres or grams to kilograms.
- Show the stages in your working. Even if your answer is wrong you can still gain method marks.
- If you have used a calculator make sure that your answer makes sense. Try doing the calculation in your head with rounded numbers.
- Give a unit with your final answer, if one is not already given.
- Be neat. Write numbers clearly. If the examiner cannot read what you have written your work will not gain credit. It may help to write a few words to explain what you have done.

How to use the 'How Science Works' snake

The snake brings together all of those ideas that you have learned about 'How Science Works'. You can join the snake at different places – an investigation might start an observation, testing might start at trial run.

How do you think you could use the snake on how marble statues wear away with acid rain? Try working through the snake using this example – then try it on other work you've carried out in class.

Remember there really is no end to the snake – when you reach the tail it is time for fresh observations. Science always builds on itself – theories are constantly improving.

OBSERVATION

HYPOTHESIS

PREDICTION

I wonder why…

Perhaps it's because…

I think that if…

I should be honest and tell it as it is. Does the data support or go against my hypothesis?

Is it a linear (straight line) relationship – positive, negative or directly proportional (starting at the origin)? or is it a curve – complex or predictable?

Which of these should I use?
- Bar chart
- Line graph
- Scatter graph

RELATIONSHIP SHOWN BY DATA

PRESENTING DATA

Am I going further than the data allows me?

Are the links I have found – causal, by association or simply by chance?

CONCLUSION

Have I given a balanced account of the results?

My conclusion would be more reliable and valid if I could find some other research to back up my results.

Just how reliable (trustworthy) was the data? Would it be more reliable if somebody else repeated the investigation? Was the data valid – did it answer the original question?

EVALUATION

USE SECONDARY DATA

There are still many questions that we cannot answer in scien

Should the variables I use be continuous (any value possible), discrete (whole number values), ordered (described in sequence) or categoric (described by words)?

I will try to keep all other variables constant, so that it is a fair test. That will help to make it valid.

DESIGN

CONTROL VARIABLE

TRIAL RUN

This will help to decide the:
- Values of the variables
- Number of repeats
- Range and interval for the variables

Can I use my prediction to decide on the variable I am going to change (independent) and the one I am going to measure (dependent)?

Are my instruments sensitive enough?

Will the method give me accuracy (i.e. data near the true value)? Will my method give me enough precision and reliability (i.e. data with consistent repeat readings)?

I'll try to keep random errors to a minimum or my results will not be precise. I must be careful!

PREPARE A TABLE FOR THE RESULTS

Are there any systematic errors? Are my results consistently high or low?

CARRY OUT PROCEDURE

Are there any anomalies (data that doesn't follow the pattern)? If so they must be checked to see if they are a possible new observation. If not, the reading must be repeated and discarded if necessary.

I should be concerned about the ethical, social, economic and environmental issues that might come from this research.

I should be careful with this information. This experimenter might have been biased – must check who they worked for; could there be any political reason for them not telling the whole truth? Are they well qualified to make their judgement? Has the experimenter's status influenced the information?

The final decisions should be made by individuals as part of society in general.

Could anyone exploit this scientific knowledge or technological development?

TECHNOLOGICAL DEVELOPMENTS

There are questions that science cannot answer at all – such as 'Should we...?' questions.

C1a | Products from rocks

Checklist

This spider diagram shows the topics in the unit. You can copy it out and add your notes and questions around it, or cross off each section when you feel confident you know it for your exams.

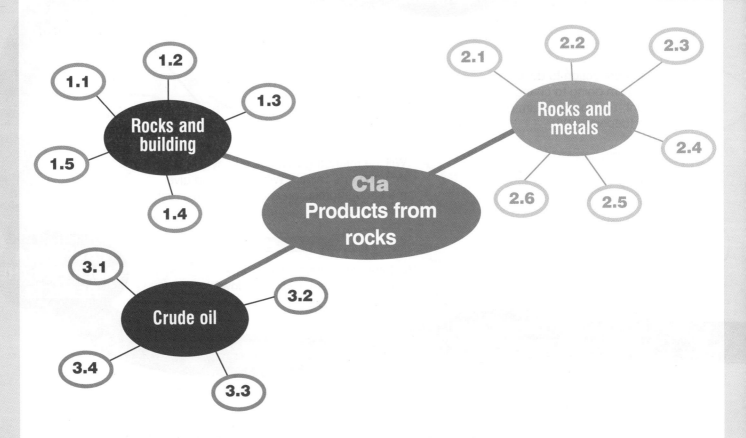

<table>
<tr><td colspan="4">

Tick when you:

</td></tr>
<tr><td>reviewed it after your lesson</td><td>☑</td><td>☐</td><td>☐</td></tr>
<tr><td>revised once – some questions right</td><td>☑</td><td>☑</td><td>☐</td></tr>
<tr><td>revised twice – all questions right</td><td>☑</td><td>☑</td><td>☑</td></tr>
<tr><td colspan="4">

Move on to another topic when you have all three ticks.

</td></tr>
</table>

Chapter 1 Rocks and building

1.1	Atoms, elements and compounds	☐ ☐ ☐
1.2	Limestone and its uses	☐ ☐ ☐
1.3	Decomposing carbonates	☐ ☐ ☐
1.4	Quicklime and slaked lime	☐ ☐ ☐
1.5	Cement, concrete and glass	☐ ☐ ☐

Chapter 2 Rocks and metals

2.1	Extracting metals	☐ ☐ ☐
2.2	Extracting iron	☐ ☐ ☐
2.3	Properties of iron and steels	☐ ☐ ☐
2.4	Alloys in everyday use	☐ ☐ ☐
2.5	Transition metals	☐ ☐ ☐
2.6	Aluminium and titanium	☐ ☐ ☐

Chapter 3 Crude oil

3.1	Fuels from crude oil	☐ ☐ ☐
3.2	Fractional distillation	☐ ☐ ☐
3.3	Burning fuels	☐ ☐ ☐
3.4	Cleaner fuels	☐ ☐ ☐

What are you expected to know?

Chapter 1 Rocks and building (See students' book pages 24–35)

- All substances are made of atoms that have a tiny central nucleus surrounded by electrons.
- The simplest substances contain only one type of atom and are called 'elements'.
- Atoms of each element have their own chemical symbol.
- The periodic table lists all the known elements in groups with similar chemical properties.
- Elements form compounds by combining with other elements: their atoms give, take or share electrons, so they bond together.
- The formula of a substance shows the types of atom it contains and how many have combined together.
- In chemical reactions, the mass of the reactants is always the same as the mass of the products.
- Chemical equations are balanced. There is the same number of each type of atom on each side.
- Limestone is used as a building material and to make quicklime, slaked lime, cement and glass.
- Heating metal carbonates produces metal oxides and carbon dioxide gas.

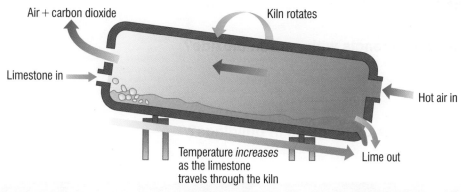

A rotary lime kiln

Chapter 2 Rocks and metals (See students' book pages 38–51)

- We get metals from ores that are dug from the Earth.
- Most metals have to be extracted from their ores using chemical reactions, and this can use large amounts of energy.
- A metal mixed with other elements is called an alloy.
- Alloys have different properties to pure metals and can be designed for specific uses.
- Iron, copper, gold, aluminium, titanium and their alloys have many uses.
- Recycling, and new ways of extracting metals, helps to conserve energy and resources.

Chapter 3 Crude oil (See students' book pages 54–63)

- Crude oil is a mixture of many different hydrocarbons.
- Alkanes are saturated hydrocarbons with the general formula C_nH_{2n+2}
- Fractional distillation separates crude oil into fractions, many of which are used as fuels.
- Burning fuels pollutes the atmosphere with carbon dioxide, sulfur dioxide and particulates.
- Producing cleaner and better fuels can help reduce pollution.

(1) **Is the substance represented by the formula CH$_4$ an element or a compound?**

(2) **What happens to elements in chemical reactions?**

(3) **What are the main uses of limestone?**

(4) **What are the products when limestone is heated strongly?**

(5) **Which metal carbonates decompose on heating?**

(6) **Write a balanced equation for the thermal decomposition of magnesium carbonate.**

(7) **How is slaked lime made?**

(8) **What happens when slaked lime reacts with carbon dioxide?**

(9) **How is concrete made?**

(10) **What are the two main reasons for using glass in buildings?**

---

students' book
page 24

C1a 1.1   Atoms, elements and compounds

KEY POINTS

1 All substances are made of atoms.
2 Elements are substances made of only one type of atom.
3 Symbols are used for atoms to show what happens in chemical reactions.

- There are about 100 different elements from which all substances are made.
- Elements are listed in the periodic table.
- Each element is made of one type of atom.
- Atoms are represented by a chemical symbol, e.g. Na for an atom of sodium, O for an atom of oxygen.
- Atoms have a tiny nucleus surrounded by electrons. When elements react their atoms join with atoms of other elements.
- Atoms form chemical bonds by losing, gaining or sharing a small number of electrons.
- Compounds contain the atoms of two or more elements bonded together.

Key words: atom, compound, electron, element, symbol

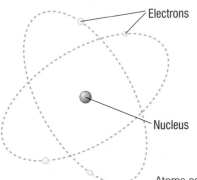

Electrons

Nucleus

Atoms consist of a small nucleus surrounded by electrons

AQA EXAMINER SAYS...

Remember that a symbol represents one atom of an element.

CHECK YOURSELF

1 What substances are made of only one type of atom?

2 What do symbols represent?

3 What is a compound?

C1a 1.2 Limestone and its uses

KEY POINTS

1 Limestone is used as a building material and to make quicklime, cement and glass.
2 Thermal decomposition of limestone makes quicklime and carbon dioxide.
3 The formula of a substance shows the number of atoms of each element that have joined together.

GET IT RIGHT!

Thermal decomposition means 'breaking down using heat': you need to make both points to get full marks.

We quarry large amounts of limestone rock because it has many uses.

Blocks of limestone can be used for building, and it is used to make:

● quicklime
● cement
● glass.

Limestone is made up mainly of calcium carbonate, formula $CaCO_3$.

When heated strongly, calcium carbonate decomposes to make quicklime (calcium oxide, CaO) and carbon dioxide (CO_2).

We can represent the reaction by the word equation:

$$calcium\ carbonate \rightarrow calcium\ oxide + carbon\ dioxide$$

This is done on a large scale in lime kilns.

Key words: formula, quicklime, thermal decomposition, word equation

CHECK YOURSELF

1 Give four uses of limestone.
2 What do we mean by 'thermal decomposition'?
3 How many different elements are there in calcium carbonate, $CaCO_3$?
4 How many atoms are there in the formula of calcium carbonate, $CaCO_3$?

C1a 1.3 Decomposing carbonates

KEY POINTS

1 Carbonates of metals decompose when heated to produce the metal oxide and carbon dioxide.
2 We can represent chemical reactions by balanced equations.
3 In a chemical reaction the mass of the reactants is equal to the mass of the products.

BUMP UP YOUR GRADE

For grade B and above: make sure you can write balanced symbol equations as well as word equations for all the reactions in this section.

Metal carbonates decompose in a similar way to calcium carbonate when heated strongly enough.

A Bunsen burner flame is not hot enough to decompose sodium carbonate or potassium carbonate.

The reactions can be represented by balanced chemical equations. For example:

$$CaCO_3 \rightarrow CaO + CO_2$$

● Symbol equations should be balanced with the same number of each type of atom on both sides.
● Atoms are not created or destroyed in chemical reactions. So the mass of the products is always the same as the mass of the reactants.

Key words: balanced chemical equation

 EXAMINER SAYS...

You should be able to write equations for the decomposition of any metal carbonate.

CHECK YOURSELF

1 Name the products when zinc carbonate is thermally decomposed.
2 Name two carbonates that do not decompose when heated with a Bunsen burner.
3 Write a balanced chemical equation for the thermal decomposition of copper(II) carbonate.
4 10.0 g of magnesium carbonate was heated strongly. What mass of products would be formed?

Quicklime and slaked lime

EXAMINER SAYS...

Make sure you can write equations for the reactions of quicklime with water and slaked lime with carbon dioxide.

EXAM HINTS

Make sure you check that any equations you write are correctly balanced – count the atoms on both sides.

- Quicklime (calcium oxide) reacts with water to produce slaked lime (calcium hydroxide).
- Calcium hydroxide is only slightly soluble in water, but a little dissolves to form a solution called lime water.
- Calcium hydroxide reacts with carbon dioxide to form calcium carbonate, which is insoluble in water. This is why carbon dioxide bubbled into lime water makes it go cloudy.
- Mortar is used to hold stone or bricks together in buildings. Lime mortar is made by mixing slaked lime with sand and adding water.
- When carbon dioxide in the air reacts with the calcium hydroxide in the mortar it forms calcium carbonate and sets hard.

Key words: lime water, mortar, slaked lime

Lime mortar should be used to repair old buildings

CHECK YOURSELF

1 Write the word equation for the reaction of carbon dioxide with calcium hydroxide.

2 Write the balanced chemical equation for carbon dioxide reacting with lime water.

3 Why does carbon dioxide turn lime water milky?

4 Why does lime mortar set hard?

Cement, concrete and glass

Glass can produce some spectacular buildings

- To make cement, limestone is mixed with clay, heated strongly and the product is powdered.
- Mortar made with cement and sand is stronger and sets faster than lime mortar. It will set in wet conditions, even under water.
- A mixture of cement, sand, stones or crushed rock and water is called 'concrete'. This can be poured into moulds or spread out before it sets to produce different shapes. Concrete can be reinforced by pouring it around steel.
- Glass is used to allow light into buildings and make them weatherproof.
- The properties of these materials can be modified for specific uses. This can be done by using different proportions of the main ingredients or by adding other substances.

Key words: cement, concrete, glass

GET IT RIGHT!

Make sure you know the differences between cement, mortar and concrete.

CHECK YOURSELF

1 Why is cement mortar more useful than lime mortar?
2 What is concrete?
3 How can concrete be made even stronger?
4 Why is glass so useful in buildings?

C1a 1   End of chapter questions

1 **Which of these substances are compounds? Ca, Cl₂, CO₂, MgCO₃, O₂**

2 **How do atoms join together in a compound?**

3 **Give four uses of limestone.**

4 **Write a word equation for the thermal decomposition of calcium carbonate.**

5 **Why is the mass of the products of a chemical reaction always the same as the mass of reactants?**

6 **Name a metal carbonate that does not decompose when heated with a Bunsen burner flame.**

7 **Write a balanced equation for the reaction of calcium hydroxide with carbon dioxide.**

8 **How is lime mortar made?**

9 **How is cement made?**

10 **Suggest two methods that could be used when making glass to change its properties.**

Pre Test: Rocks and metals

1. What is an ore?
2. What chemical reactions do we use to extract metals?
3. How is iron extracted?
4. What is cast iron?
5. Why is pure iron soft and easily bent?
6. What are the main types of steel?
7. What are alloys?
8. What are 'smart' alloys?
9. What are the properties of transition metals?
10. How is copper produced?
11. What two properties of aluminium and titanium make them especially useful metals?
12. Why is it expensive to extract aluminium and titanium?

---

students' book
page 38

C1a 2.1 Extracting metals

KEY POINTS

1. An ore contains enough metal to make it worth extracting the metal.
2. The method we use to extract a metal depends on its reactivity.
3. Many metals can be extracted from their oxides using carbon.

Rocks that contain enough of a metal or a metal compound to make it worth extracting the metal are called 'ores'.

A few very unreactive metals like gold are found native as the metal. Gold can be separated from rocks by physical methods. However, most metals are found as compounds and so have to be extracted by chemical reactions.

- Metals can be extracted from compounds by displacement using a more reactive element or by electrolysis.
- Metals that are less reactive than carbon can be extracted by heating with carbon to reduce their oxides. (Reduction is the removal of oxygen from a compound.) This method is used commercially if possible.

Key words: displacement, electrolysis, ore, native, reduce

CHECK YOURSELF

1. What are ores?
2. What is meant by the term 'native metal'?
3. Name two metals that can be extracted by reduction with carbon.
4. Name a metal that is too reactive to be extracted using carbon.
5. What happens in a reduction reaction?

C1a 2.2 Extracting iron

KEY POINTS

1 Iron ore contains iron oxide.
2 Iron oxide is reduced in a blast furnace using coke.
3 The iron that is produced is hard and brittle because it contains impurities.

AQA EXAMINER SAYS…

You do not need to remember technical details of the blast furnace.

Many of the ores used to produce iron contain iron(III) oxide.
Iron is less reactive than carbon and so it can be extracted from its ore using carbon.

Iron is extracted in a blast furnace using coke to provide the carbon. The iron oxide is reduced at high temperatures by the carbon.

The iron produced contains about 4% impurities that make it hard and brittle. Its properties mean that it has only a small number of uses as cast iron. However, it is the starting material for making steels.

Key words: coke, reduced, cast iron

CHECK YOURSELF

1 Name the compound of iron that is reduced in the blast furnace.

2 What substance provides the carbon in the blast furnace?

3 Write a word equation for a reaction that produces iron in the blast furnace.

4 Why is cast iron only useful as a metal for some purposes?

C1a 2.3 Properties of iron and steels

KEY POINTS

1 Pure iron is a soft metal that bends easily.
2 Steels are alloys of iron.
3 There are many different steels that have properties suitable for particular uses.

AQA EXAMINER SAYS…

Make sure you know the three main types of steel and what they contain.

- **Pure iron** has a regular arrangement of atoms. The atoms are in layers that slide easily over each other so its shape can be easily changed.
- **Steels** are alloys of iron – they are mixtures that contain other elements as well as iron. The other elements change the regular structure of the metal and this changes its properties.
- **Carbon steels** contain small amounts of carbon, up to 1.5%. Increasing the amount of carbon in a steel makes it harder but more brittle.
- **Low-alloy steels** contain up to 5% of other metals to give them special properties.
- **High-alloy steels** contain higher percentages of other metals, e.g. stainless steel contains about 25% chromium and nickel, and it does not rust.

Key words: alloy, carbon steels, low-alloy steels, high-alloy steels

CHECK YOURSELF

1 Why is pure iron soft?

2 What are steels?

3 What is the maximum percentage of carbon in carbon steels?

4 A tall television transmitter mast is made of steel that contains 2% manganese. What type of steel is it?

Iron

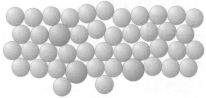

Alloy

The atoms in pure iron are arranged in layers which can easily slide over one another. In alloys the layers cannot slide so easily because atoms of other elements change the regular structure.

C1a 2.4 Alloys in everyday use

KEY POINTS

1 A pure metal is alloyed by mixing it with other metals or elements.

2 Alloying changes the properties of a pure metal so that it is more useful.

3 Smart alloys have special properties such as returning to their original shape.

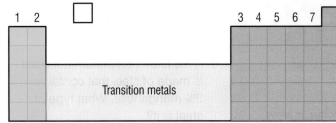

This dental brace pulls the teeth into the right position as it warms up. That's smart!

- Many pure metals are too soft and lose their shape too easily. Mixing a pure metal with other metals or elements to make an alloy makes the metal harder.
- Alloying may also affect other properties of the metal, such as strength, appearance and resistance to corrosion. The cost of an alloy depends on the cost of the metals it contains.
- Alloys are mixtures. This means that the amounts of the elements they contain can be varied so they can be designed with specific properties for a particular use.
- Shape memory alloys are 'smart' because they return to their original shape after they have been bent.

Key words: alloy, mixture, shape memory alloys

AQA↗ EXAMINER SAYS...

When asked for properties, many students include cost or cheapness but cost is not a property of a substance.

CHECK YOURSELF

1 List four properties of a metal that can be changed by alloying.

2 Why can the amounts of the elements in an alloy be varied?

3 Why are some alloys 'smart'?

C1a 2.5 Transition metals

KEY POINTS

1 Transition metals have many similar properties that make them very useful.

2 Copper is a very good conductor of heat and electricity that does not corrode easily.

3 Extracting copper by traditional methods is expensive and affects the environment.

The elements between Groups 2 and 3 in the periodic table are all metals and are called the 'transition metals'.

- They are good conductors of heat and electricity.
- They are strong, hard and dense, but can be bent or hammered into shape.
- Except for mercury, transition metals have high melting points.

Transition metals have many similar properties but there are differences that make them useful for specific purposes.

Copper is particularly useful for plumbing and electrical wires. Copper is processed by smelting and electrolysis.

Little high-grade copper ore remains, and so huge amounts of rock have to be moved in open cast mines. New methods of extracting copper are being developed, including using bacteria, fungi and plants.

Key words: transition metals, smelting, electrolysis

CHECK YOURSELF

1 Where in the periodic table are the transition elements?

2 In what ways are the transition elements similar?

3 Why are new methods to extract copper being developed?

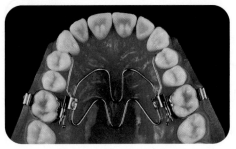

| 1 | 2 | | | | | | | 3 | 4 | 5 | 6 | 7 | |

Transition metals

The transition metals

C1a 2.6 Aluminium and titanium

KEY POINTS

1 Aluminium and titanium have low densities compared with many other metals.
2 They also resist corrosion, and titanium is strong at high temperatures.
3 Extracting these metals is expensive because it involves electrolysis and high temperatures.
4 Recycling saves resources and the energy needed to extract the metal from its ore.

We use aluminium and titanium where low density and resistance to corrosion are important. Both metals form oxide layers that protect them from further corrosion.

- Aluminium is soft with quite a low melting point but can be hardened by alloying.
- Titanium reacts with carbon and so is extracted by displacement with a reactive metal e.g. sodium.
- Aluminium and sodium are too reactive to extract with carbon and so electrolysis is used, with high costs for energy.
- Aluminium is widely used in buildings, cans, cooking foil, electricity cables and aircraft.
- Titanium is used in jet engines, nuclear reactors, replacement hip joints and bicycles.

Recycling metals avoids mining and processing metal ores.

Key words: displacement, oxide layers, recycling

AQA EXAMINER SAYS…

As well as knowing the similarities between aluminium and titanium, make sure you know the differences.

GET IT RIGHT!

Aluminium is a reactive metal, near to the top of the reactivity series.

However, it is protected by a tough layer of oxide on its surface.

CHECK YOURSELF

1 What properties make aluminium and titanium especially useful metals?
2 Why is titanium extracted by displacement using a very reactive metal?
3 Why is aluminium extracted using electrolysis?
4 What are the benefits of recycling metals?

C1a 2   End of chapter questions

1 How much metal or metal compound is in an ore?

2 Suggest how zinc would be extracted from zinc oxide commercially.

3 Why does the reaction to produce iron involve reduction?

4 Why is cast iron brittle?

5 Name two elements in all carbon steels.

6 What is meant by 'high-alloy steels'?

7 Why are many metals used as alloys?

8 What type of alloys return to their original shape?

9 All transition metals conduct heat and electricity. Give one other property of all transition metals (except mercury).

10 Give one advantage of using bacteria or fungi to help extract copper.

11 What protects aluminium and tin from corrosion?

12 How is electrolysis involved in the extraction of titanium?

1. What two elements make up the main compounds in crude oil?

2. What are alkanes?

3. What method is used to separate crude oil in a refinery?

4. What are fractions?

5. What are the products of burning hydrocarbons completely?

6. What other substances can be produced by burning fossil fuels?

7. What are the three main effects on the environment of burning fossil fuels?

8. How can we reduce the effects of burning fossil fuels?

---

students' book
page 54

C1a 3.1 Fuels from crude oil

KEY POINTS

1. Crude oil is a mixture of many different compounds.
2. Distillation separates liquids with different boiling points.
3. Most of the compounds in crude oil are hydrocarbons, and many of these are alkanes.

Crude oil contains many different compounds that boil at different temperatures. These burn under different conditions so it needs to be separated to make useful fuels.

We can separate mixtures of liquids by distillation. Simple distillation of crude oil can produce liquids that boil within different temperature ranges.

Most of the compounds in crude oil are hydrocarbons. This means that they contain carbon and hydrogen only. Many of these hydrocarbons are alkanes, with the general formula C_nH_{2n+2}. Alkanes contain as many hydrogen atoms as possible in each molecule and so we call them saturated hydrocarbons.

Molecules can be represented by:

- a molecular formula that shows the *number* of each type of atom
- a structural formula that shows *how* the atoms are bonded together.

Key words: distillation, hydrocarbons, alkanes, molecular formula, saturated hydrocarbons, structural formula

Ethane

Propane

We can represent alkanes like this, showing all of the atoms in the molecule. The line between two atoms in the molecule is the chemical bond holding them together.

CHECK YOURSELF

1. Why does crude oil need to be separated?

2. What method can we use to separate mixtures of liquids?

3. What are hydrocarbons?

4. How can we represent molecules?

C1a 3.2 Fractional distillation

Crude oil is separated at refineries by fractional distillation.

The crude oil is vaporised and fed into a fractionating column. This is a tall tower that is hot at the bottom and cooler at the top.

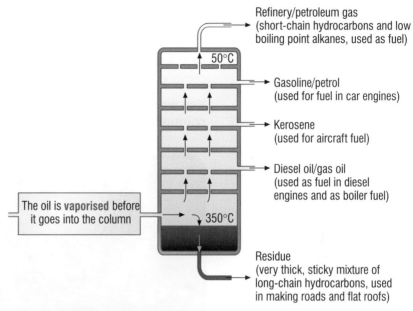

Refinery/petroleum gas
(short-chain hydrocarbons and low boiling point alkanes, used as fuel)

Gasoline/petrol
(used for fuel in car engines)

Kerosene
(used for aircraft fuel)

Diesel oil/gas oil
(used as fuel in diesel engines and as boiler fuel)

The oil is **vaporised** before it goes into the column

50°C

350°C

Residue
(very thick, sticky mixture of long-chain hydrocarbons, used in making roads and flat roofs)

We use fractional distillation to turn the mixture of hydrocarbons in crude oil into fractions, each containing compounds with similar boiling points

Inside the column there are many trays with holes to allow gases through. The vapours move up the column and condense on the trays when they reach their boiling points.

There are outlets at different levels to collect the liquid fractions.

Hydrocarbons with the *smallest* molecules have the *lowest* boiling points and so are collected at the *top* of the tower. The fractions collected at the bottom have the highest boiling points.

Fractions with low boiling points burn more easily, which makes them more useful as fuels.

Key words: fractionating column, fractions

EXAMINER SAYS...

Some students confuse simple distillation done in the laboratory with fractional distillation. Simple distillation is done in *steps*, but fractional distillation is a *continuous process*.

CHECK YOURSELF

1 Name the process used to separate crude oil in a refinery.

2 How does this process work?

3 What sort of hydrocarbons are in the fractions collected near the top of the column?

4 Which fractions are more difficult to burn?

KEY POINTS

1 Burning hydrocarbons in plenty of air produces carbon dioxide and water.
2 In a limited supply of air, carbon monoxide and particles may be produced.
3 Any sulfur compounds in the fuel burn to produce sulfur dioxide.

When pure hydrocarbons burn completely, they produce carbon dioxide and water. However, the fuels we use are not always burnt completely and they may also contain other substances.

In a limited supply of air, such as in an engine, carbon monoxide and carbon may also be produced, and some of the hydrocarbons may not burn. Carbon and unburnt hydrocarbons form tiny particles in the air.

Most fossil fuels contain sulfur compounds. When the fuel burns these sulfur compounds produce sulfur dioxide.

Key words: carbon monoxide, unburnt hydrocarbons, particles, sulfur compounds, sulfur dioxide

The effect of many cars in a small area. Smog formed from car pollution can harm us.

BUMP UP YOUR GRADE

For grade C and above, you should be able to write balanced symbol equations for the combustion of an alkane, given its formula.

AQA EXAMINER SAYS...

You should be able to write equations for burning hydrocarbons.

CHECK YOURSELF

1 Name the products when hydrocarbons burn in plenty of air.

2 What else may be produced in a limited supply of air?

3 Why is sulfur dioxide produced when fossil fuels burn?

C1a 3.4 Cleaner fuels

KEY POINTS

1 Carbon dioxide contributes to global warming.
2 Particulates cause global dimming.
3 Sulfur dioxide forms acid rain.

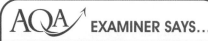

EXAMINER SAYS...

You should be able to link the products of burning fossil fuels with three main environmental effects. Other effects are covered in B1b, Chapter 8 'How people affect the planet'. You should also know about some developments to overcome problems with fuels.

The products from burning fossil fuels are released into the atmosphere in exhaust gases.

- Carbon dioxide is a greenhouse gas. It reduces the amount of heat lost from the Earth and causes global warming.
- Particulates cause health problems and cause global dimming by reflecting sunlight away from the Earth.
- Sulfur dioxide dissolves in water and forms acid rain. The amount of sulfur dioxide released can be reduced by removing sulfur compounds from fuels at the refinery, or by removing sulfur dioxide from the waste gases after burning.

Fossil fuel resources are being used up and will run out in the future. Plants can provide sugars to make ethanol, or can produce oils that can be used as biodiesel.

Key words: greenhouse gas, global warming, global dimming, acid rain

We can use plants that make sugar to produce ethanol by fermenting the sugar using yeast. We can then add the ethanol to petrol, making **gasohol**. Not only does this reduce the amount of oil needed, it also produces less pollution because gasohol burns more cleanly than pure petrol.

CHECK YOURSELF

1 How does burning fossil fuels cause global warming?

2 What is 'global dimming'?

3 How can we reduce the amount of sulfur dioxide released?

C1a 3 End of chapter questions

1 **Why can crude oil be separated by distillation?**

2 **What is the formula of the alkane with four carbon atoms?**

3 **What are the two main processes in fractional distillation?**

4 **Where are the fractions that burn most easily collected?**

5 **Write a word equation for the complete combustion of the hydrocarbon called propane.**

6 **What is in the particles produced by incomplete combustion of fossil fuels?**

7 **Why does burning fossil fuels contribute to global warming?**

8 **Name the element in compounds in fossil fuels that results in acid rain.**

1 Read the passage and use the information to help you answer the questions.

> An open cast mine at Bingham Canyon provides over 10% of the copper used by the USA. Bingham Canyon is in mountains at the edge of the Great Salt Lake Desert. In the past, miners at Bingham Canyon dug underground mines into a mountain rich in metal ores. They were mining silver and gold and ignored the copper ore. When the gold and silver had been removed the area went into decline. A mining company found that the mountain contained large deposits of low-grade copper ore and decided to extract it using an open cast mine. Every day the mine produces 685 tonnes of copper by digging 225 000 tonnes of rock. The mountain has now been removed and replaced by a huge open pit. The waste is dumped in the canyon and in a salt lake that was used by birds and other wildlife. The mine attracted thousands of migrant workers to the area, and the mining company built new towns to replace the old temporary towns that the underground miners had put up.

(a) What first brought miners to Bingham Canyon?
(1 mark)

(b) What is the percentage of copper in the ore according to the figures in the passage? (1 mark)

(c) Why did the underground miners in the past ignore the copper ore? (1 mark)

(d) Suggest why an open cast mine is used instead of an underground mine to extract the copper ore.
(2 marks)

(e) Describe the advantages and disadvantages of the mine at Bingham Canyon. (5 marks)

2 The table shows the percentages of elements in some different types of iron and steel:

Type of iron or steel	Fe	C	Si	P	Mn	Cr	Ni
Iron from blast furnace	94.50	4.40	0.70	0.10	0.40	0	0
Mild steel	99.90	0.04	0.04	0.01	0.02	0	0
Tool steel	94.33	0.75	0.40	0.02	0.40	4.00	0.10
Stainless steel	70.53	0.03	0.20	0.04	0.50	17.2	11.5

(a) Name the compound of iron that is reduced in a blast furnace. (1 mark)

(b) Name the reducing agent used in the blast furnace. (1 mark)

(c) What must be removed from the iron from the blast furnace to make steels? (1 mark)

(d) Which metal could be described as a low-alloy steel? (1 mark)

(e) Which steel will be the most expensive? (1 mark)

(f) Why is mild steel soft and easy to shape? (2 marks)

3 Crude oil can be separated in a fractionating column.

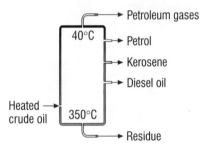

(a) Why can crude oil be separated in this way? (1 mark)

(b) Why is the crude oil heated before it goes into the column? (1 mark)

(c) Explain how the different fractions are produced.
(4 marks)

(d) Most of the molecules in the fractions are alkanes.
  (i) What is the formula of the alkane with three carbon atoms in its molecules? (1 mark)
  (ii) In which fraction would you be most likely to find this alkane? (1 mark)

(e) Diesel engines produce more carbon and unburnt hydrocarbons than petrol engines.
  (i) Why does this happen in engines that use diesel fuel? (2 marks)
  (ii) What environmental problem is caused by carbon and unburnt hydrocarbons? (1 mark)

4 Complete the sentences by using the correct word from the list:

alloys   compounds   elements   metals   mixtures

(a) Carbon, iron and oxygen are all . . . .
(b) Bronze, calcium and copper are all . . . .
(c) Concrete, mortar and steel are all . . . .
(d) Calcium carbonate, carbon dioxide and water are all . . . . (4 marks)

 Test & Assessment Interactive quizzes, answers and hints online!

The answer is worth 4 marks out of the 5 available. The responses worth a mark are underlined in red.

We can improve the answer in several ways:

Crude oil is separated into fractions in oil refineries. This is done in a tall column by fractional distillation. Explain, as fully as you can, how this method of separation works. *(5 marks)*

The crude oil is heated to vaporise it and is fed into the tall column. The column is hot at the bottom and cool at the top. The vapours rise up the column and condense at different temperatures. The fractions come out at different heights up the column. The ones with the lowest boiling points are collected at the top. These are petroleum gases and do not condense but can be liquefied for use as fuels.

'The vapours condense at different temperatures' is only partly correct. The fifth mark could be gained by stating that the fractions, or molecules or compounds, **condense when they reach their boiling point** in the column.

This sentence about uses of the gases is not relevant to the question.

'The fractions come out at different heights' is too vague. To gain a fifth mark, we could state that the **higher boiling fractions condense near the bottom** of the column and the lower boiling fractions near the top. Alternatively, we could state that the liquid fractions **condense (on plates or trays) at different levels according to their boiling points**.

The answer is worth 4 marks out of the 5 available. The responses worth a mark are underlined in red.

We can improve the answer in several ways:

Aluminium is not an unreactive metal. It is quite high in the reactivity series.

Aluminium is used to make cans, cooking foil, saucepans, high voltage cables and aircraft. Explain, as fully as you can, why aluminium is such a useful metal. *(5 marks)*

Aluminium has so many uses because it is a light and unreactive metal. Acids in drinks or food do not attack it so it can be used as cans and for cooking. It is lighter than other metals and is not corroded by air so it is used for aircraft. It is a good conductor of heat and electricity, so is used for saucepans and high voltage cables.

This answer contains some correct properties, but these are not always correctly related to its uses. Other properties that would gain marks include that it is **flexible** or **bendable** and that it can be used in **alloys**.

It is a good conductor of electricity, but that is not why it is used for saucepans. This sentence would be better as two separate statements linking **conduction of heat** with saucepans and **electricity** with high voltage cables.

'It is a light metal' is not correct, but stating it is **lighter than other metals** or it is **less dense than other metals** gains a mark.

It does not corrode in air or acids because it **is protected by an oxide layer**.

C1b | Oils, Earth and atmosphere

Checklist

This spider diagram shows the topics in the unit. You can copy it out and add your notes and questions around it, or cross off each section when you feel confident you know it for your exams.

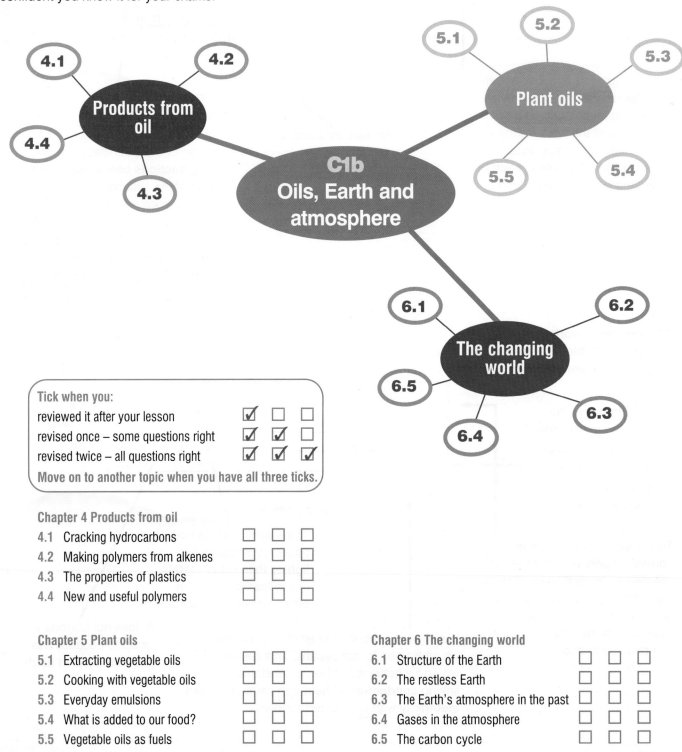

Tick when you:

reviewed it after your lesson	☑	☐	☐
revised once – some questions right	☑	☑	☐
revised twice – all questions right	☑	☑	☑

Move on to another topic when you have all three ticks.

Chapter 4 Products from oil

4.1	Cracking hydrocarbons	☐	☐	☐
4.2	Making polymers from alkenes	☐	☐	☐
4.3	The properties of plastics	☐	☐	☐
4.4	New and useful polymers	☐	☐	☐

Chapter 5 Plant oils

5.1	Extracting vegetable oils	☐	☐	☐
5.2	Cooking with vegetable oils	☐	☐	☐
5.3	Everyday emulsions	☐	☐	☐
5.4	What is added to our food?	☐	☐	☐
5.5	Vegetable oils as fuels	☐	☐	☐

Chapter 6 The changing world

6.1	Structure of the Earth	☐	☐	☐
6.2	The restless Earth	☐	☐	☐
6.3	The Earth's atmosphere in the past	☐	☐	☐
6.4	Gases in the atmosphere	☐	☐	☐
6.5	The carbon cycle	☐	☐	☐

What are you expected to know?

Chapter 4 Products from oil *See students' book pages 70–79*

- We use cracking to break down hydrocarbons into smaller molecules.

- The products of cracking include alkanes that are used as fuels and alkenes that can be used to make other chemicals and polymers.

- Alkenes are unsaturated with a general formula C_nH_{2n} and contain a double carbon–carbon bond.

- Alkene molecules are monomers that can add together to make polymers.

- Polymers with different properties can be made by using different starting materials and by using different reaction conditions. This allows new polymers to be developed for specific uses.

- Polymers that are not biodegradable can lead to problems with waste disposal.

Chapter 5 Plant oils

See students' book pages 82–93

- Vegetable oils are important as foods and fuels.

- Vegetable oils can be extracted from seeds, nuts and fruits.

- Oils do not dissolve in water, but can be mixed with water to produce emulsions that have many uses.

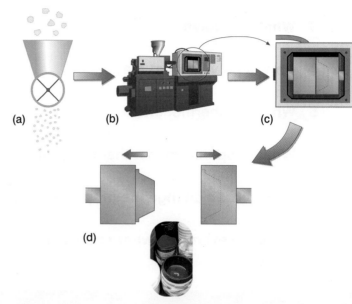

A very common way of making things out of polymers is to use a thermosoftening plastic that can be shaped in a mould: **(a)** chunks of monomer are ground into small pieces. **(b)** These are heated to melt them and then . . . **(c)** the molten plastic is forced into a mould. **(d)** The mould is separated to release the finished article.

- Many vegetable oils are unsaturated. This can be tested using bromine or iodine.

- Unsaturated oils can be made into solids at room temperature by hydrogenation to produce saturated fats.

- Chemical analysis can be used to identify the additives that may be contained, legally or illegally, in foods.

Chapter 6 The changing world *See students' book pages 96–107*

- Mountains were once thought to have formed by the Earth shrinking as it cooled.

- It is now thought that movements of large parts of the Earth's crust and upper mantle, called tectonic plates, produce mountains, volcanoes and earthquakes.

- It is believed that the oceans were formed from water vapour released by volcanoes and that the Earth's early atmosphere was mainly carbon dioxide.

- Plants took up carbon dioxide by photosynthesis, and the oceans dissolved carbon dioxide; plants produced the oxygen that is now in the atmosphere.

- The atmosphere has changed little in the last 200 million years, but recently, burning fossil fuels has increased the proportion of carbon dioxide.

1. What is cracking?

2. What are alkenes?

3. What is a polymer?

4. How is poly(ethene) made?

5. What is a thermosoftening polymer?

6. What happens when a thermosetting polymer is heated for the first time?

7. How can we change the properties of polymers?

8. Give two examples where polymers have been designed for specific uses.

C1b 4.1 | Cracking hydrocarbons

KEY POINTS

1 Cracking of fractions from crude oil produces smaller, more useful molecules.
2 Alkanes and unsaturated hydrocarbons called alkenes are produced.
3 We can use bromine water to test for unsaturation.

Ethene — double bond

Ethene has a carbon–carbon double bond in it

Fractions from crude oil can be broken down by thermal decomposition in a catalytic cracker. The fraction is vaporised and passed over a hot catalyst, which causes the molecules to split apart and form smaller molecules.

Some of the smaller molecules are alkanes, but some are alkenes that contain carbon–carbon double bonds.

- Alkenes are unsaturated, because they contain fewer hydrogen atoms than alkanes with the same number of carbon atoms.
- Their general formula is C_nH_{2n}.
- They are hydrocarbons, like alkanes, and so burn in air.
- However, they are more reactive than alkanes. So alkenes will react with bromine water turning the orange-yellow solution colourless.

Key words: catalytic cracker, alkenes, C_nH_{2n}, bromine water

GET IT RIGHT!

'Unsaturated' means that the molecule contains fewer hydrogen atoms than an alkane molecule with the same number of carbon atoms.

AQA EXAMINER SAYS...

Make sure you use the correct words to describe the results of the bromine water test – it turns colourless (colour removed) *not* clear (which just means that you can see through it).

CHECK YOURSELF

1 What type of reaction is cracking?

2 What causes the molecules to split apart?

3 What are alkenes?

4 How can you test for unsaturation?

Making polymers from alkenes

Polymers are very large molecules made from many small molecules that have joined together. The small molecules used to make polymers are called monomers.

Lots of ethene molecules can join together in long chains to form poly(ethene), commonly called polythene. The reaction is called addition polymerisation, because the molecules simply add together and only the polymer is produced.

We can react other alkenes together in a similar way to form polymers such as poly(propene). Many of the plastics we use as bags, bottles, containers and toys are made from alkenes.

Key words: monomers, polymers, addition polymerisation, poly(ethene), poly(propene)

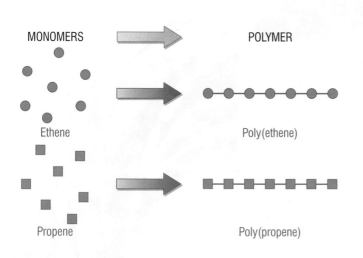

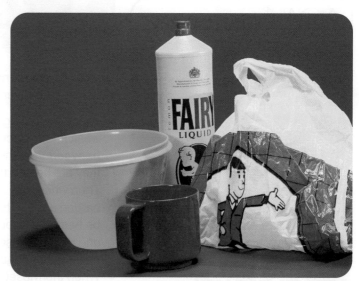

Polymers produced from oil are all around us and are part of our everyday lives

GET IT RIGHT!

In addition reactions there is only *one* product because the reactants add together.

AQA EXAMINER SAYS...

You should be able to recognise alkenes used as monomers from their names or formulae.

EXAM HINTS

Double bonds in the monomer become *single bonds* in the polymer when the molecules have joined together.

CHECK YOURSELF

1 What are monomers?

2 What is the chemical name of the polymer formed from ethene?

3 Why are polymers formed from alkenes called 'addition polymers'?

4 How are polymers made from alkenes used?

KEY POINTS

1 Polymers have very long molecules with strong bonds holding the atoms together within the molecules.
2 Thermosoftening polymers have weak forces between the polymer molecules.
3 Strong bonds form between the molecules in thermosetting polymers.

AQA EXAMINER SAYS...

Many students in their answers do not make it clear whether they are describing chemical bonds or intermolecular forces. Chemical bonds are strong and join atoms together to make molecules. Intermolecular forces act between molecules and are weaker than chemical bonds. It is best to use the word 'bond' only for forces that hold atoms together within molecules.

GET IT RIGHT!

Thermosoftening plastics become soft when heated.

Plastic kettles are made out of thermosetting plastics

Polymers have very long molecules that form a tangled mass, rather like spaghetti. This gives them strength with flexibility.

The strands of spaghetti are like the polymer molecules in a plastic

Using different monomers produces polymers with different properties.

- Thermosoftening polymers have weak intermolecular forces **between** their molecules, so when they are heated they become soft. Then when they cool down, they harden again. This means they can be heated to mould them into shape and they can also be remoulded by heating them again.
- When thermosetting polymers are heated for the first time, chemical bonds form between the polymer molecules. They link together in a giant network. These strong bonds make the plastic set hard and it cannot be softened again by heating.

The forces between the molecules in poly(ethene) are relatively weak. This means that this plastic softens fairly easily when heated.

We use thermosoftening plastics where flexibility is important and where they are not exposed to very high temperatures. Thermosetting plastics are more rigid and can withstand higher temperatures.

Key words: thermosoftening, thermosetting

CHECK YOURSELF

1 Why are polymers strong and flexible?

2 Why can a thermosoftening plastic be remoulded?

3 Why do thermosetting plastics set hard when heated?

4 What type of plastic would be best to make a garden hosepipe?

New and useful polymers

EXAMINER SAYS...

You do not need to remember the names and technical details of specific polymers. However, make sure you can recognise the types of polymer from a description of their properties or uses.

EXAM HINTS

You should be able to explain why polymers have replaced other materials.

We can design the properties of polymers by choosing different monomers and by changing the conditions used to make them.

Polymers are widely used for food packaging to keep food in good condition. Some of these polymers are not biodegradable and cause problems with waste disposal.

The polymers used for drinks bottles are strong, flexible, lightweight, clear and non-porous.

Polymers have been developed to coat fabrics that make them waterproof but able to let gases through (breathable). New polymers have been developed for medical use, including hydrogels which are also used in agriculture and food. Smart polymers can be used to control the release of drugs and shape memory polymers are used for stitching wounds.

Key words: hydrogels, smart polymers, shape memory polymers

CHECK YOURSELF

1 How can we make polymers with different properties?

2 Why are polymers widely used in food packaging?

3 How do new polymers help people keep comfortable when they are outdoors in wet weather?

4 Give one example of a new type of polymer designed for medical use.

When a shape memory polymer is used to stitch a wound loosely, the temperature of the body makes the thread tighten and close the wound, applying just the right amount of force. Later, after the wound is healed, the material is designed to dissolve and is harmlessly absorbed by the body.

C1b 4   End of chapter questions

1 **Why are fractions from crude oil cracked?**

2 **How are alkenes different from alkanes with the same number of carbon atoms?**

3 **Why is the reaction used to make poly(ethene) called addition polymerisation?**

4 **Name an alkene that can be used to make a polymer other than poly(ethene).**

5 **What type of force is between the polymer chains in a thermosoftening plastic?**

6 **Why is it not possible to remould a thermosetting plastic?**

7 **Suggest two reasons why drinks bottles are often made of plastic rather than glass.**

8 **What type of new polymers have many uses in medicine, agriculture and food?**

(1) What are the two main uses of vegetable oils?

(2) How are vegetable oils extracted?

(3) Why is fast food often cooked in oil?

(4) What has been done to a hydrogenated oil?

(5) What is an emulsion?

(6) Why are emulsifiers added to some foods?

(7) What is a food additive?

(8) What are the E numbers on food labels?

(9) What is biodiesel?

(10) Why is biodiesel a good alternative to fossil fuels?

---

students' book page 82

C1b 5.1   Extracting vegetable oils

KEY POINTS

1 Vegetable oils can be extracted from seeds, nuts and fruits by pressing or by distillation.
2 Vegetable oils provide a lot of energy as foods or fuels.
3 Some vegetable oils are unsaturated because their molecules contain carbon–carbon double bonds.

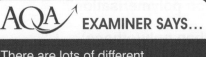

AQA EXAMINER SAYS...

There are lots of different vegetable oils, but they all have molecules with chains of carbon atoms.

Some seeds, nuts and fruits are rich in vegetable oils. The oils can be extracted by pressing followed by removing water and other impurities. Some oils are extracted by distilling the plants mixed with water. This produces a mixture of oil and water from which the oil can be separated.

The molecules in vegetable oils have hydrocarbon chains. Those with carbon–carbon double bonds are unsaturated. If there are several double bonds in each molecule, they are called polyunsaturated. Unsaturated oils will react with bromine or iodine. Bromine water is used as the test for an unsaturated compound.

Vegetable oils produce a lot of energy when eaten or when we burn them as fuels.

Key words: pressing, distilling, unsaturated, decolourise

Crushing olives before extracting the oil

CHECK YOURSELF

1 What two methods are used to extract vegetable oils?

2 Why are vegetable oils important as foods *and* fuels?

3 What is meant by an unsaturated oil?

4 What can you use to test an oil for unsaturation?

C1b 5.2 Cooking with vegetable oils

 EXAMINER SAYS…

Increasing the temperature makes chemical reactions go faster, so food cooks faster in oil than in water.

GET IT RIGHT!

Oils are liquid at room temperature, fats are solid.

The boiling points of vegetable oils are higher than water, so food is cooked at higher temperatures in oil. This means it cooks faster. It also changes the flavour, colour and texture of the food. Some of the oil is absorbed and so the energy content of the food increases.

Unsaturated oils can be reacted with hydrogen so that some or all of the double carbon–carbon bonds become single bonds. This is an addition reaction called hydrogenation and is done at about 60°C using a nickel catalyst. Hydrogenation is used to increase the melting points of oils so they harden and become solid fats at room temperature.

Solid fats can be spread and can be used to make cakes, biscuits and pastries.

Key word: hydrogenation

CHECK YOURSELF

1 Why does food cook faster in hot oil than in boiling water?
2 In what four ways is food cooked in oil different from food cooked in water?
3 What type of chemical reaction is hydrogenation?
4 Why are oils hydrogenated?

C1b 5.3 Everyday emulsions

Emulsions are made from liquids that usually separate from each other. They are made by vigorously shaking, stirring or beating the liquids together to form tiny droplets of the liquids. The droplets are so small that they remain suspended in each other and are slow to separate.

Emulsifiers help keep the droplets to stay suspended and stop the liquids from separating. They do this because different parts of their molecules are attracted to the different liquids.

Emulsions are opaque and usually thicker than the liquids they are made from. This improves their texture, appearance and ability to stick to solids.

Milk, sauces, salad dressings and ice cream are examples of emulsions.

Key words: emulsifiers

GET IT RIGHT!

Liquids that do not mix and usually separate from each other can be made into emulsions.

EXAMINER SAYS…

Emulsions are different from solutions. In a solution the substances mix completely and the liquid becomes clear. In an emulsion the liquids remain as tiny droplets and the mixture is not transparent.

CHECK YOURSELF

1 How are emulsions made?
2 How do emulsifiers work?
3 How is an emulsion different from the liquids it is made from?

What is added to our food?

Substances added to food to improve its appearance, flavour, texture and keeping qualities are called additives. Additives may be natural products or synthetic chemicals.

Some substances, like salt, vinegar and sugar have been used for hundreds of years.

There are six main types of additive:

- colours
- preservatives
- antioxidants
- emulsifiers
- acidity regulators
- flavourings.

In the European Union only permitted substances may be added to food and these are given E-numbers. Additives must be included in the list of ingredients on food labels, and can be labelled with their full name or their E-number.

Foods are checked by chemical analysis to ensure only permitted additives have been used. The methods used include chromatography and mass spectrometry.

Key words: additives, E-numbers, chromatography, mass spectrometry

AND TOMATO FLAVOUR SAUCE *AND A SACHET OF TOMATO SAUCE*

Ingredients: Wheatflour, Vegetable Oil with Antioxidants (E320, E321), Cheese & Tomato Flavour [Flavour Enhancer (621), Flavouring, Colours (E102, E110, E124, 154), Acidity Regulators (E262, E331), Acetic Acid, Citric Acid, Artificial Sweetener (Saccharin)], Maltodextrin, Salt, Tomato, Sweetcorn, Chives, Preservative (E220), Sachet: Tomato Sauce.

CHECK YOURSELF

1 What are the four main reasons for adding substances to food?

2 Why are some substances given E-numbers?

3 How could you tell from a label that a food has additives?

4 What can be done to check if substances have been added to a food?

C1b 5.5 Vegetable oils as fuels

KEY POINTS

1 We can burn vegetable oils to produce energy.
2 Modified vegetable oils can replace some of the fossil fuels we use.
3 Biodiesel is renewable and causes less pollution than fossil fuels.

Vegetable oils produce a lot of energy when they burn. They can be treated to remove some chemicals and then used as fuel in diesel engines. We can use waste vegetable oils from food frying as well as fresh oils.

Biodiesel can replace some or all of the diesel fuel produced from crude oil. Biodiesel is renewable because plants are grown to produce the vegetable oils. The plant material left after removing the oil can be used as food for animals.

Biodiesel is less harmful to the environment than fossil fuels. The plants remove carbon dioxide from the air as they grow and so when we burn it there is no additional carbon dioxide released. Biodiesel produces no sulfur dioxide and it is more biodegradable than diesel oil.

Key words: biodiesel, biodegradable, renewable

CHECK YOURSELF

1 Why is biodiesel renewable?

2 How does biodiesel help with disposal of food waste?

3 Why is biodiesel less harmful to the environment than fossil fuels?

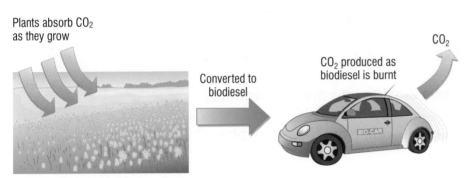

Plants absorb CO_2 as they grow

Converted to biodiesel

CO_2 produced as biodiesel is burnt

CO_2

BIO-CAR

Cars run on biodiesel produce no additional CO_2 overall, as CO_2 is absorbed by the plants which are the raw material for the fuel

C1b 5   End of chapter questions

1 **Describe the method to extract vegetable oils without using heat.**

2 **How could you test an oil for unsaturation?**

3 **Why does the energy content of food cooked in oil increase?**

4 **What is done to harden vegetable oils?**

5 **What sort of substances are used to make an emulsion?**

6 **Give two examples of foods that are emulsions.**

7 **What type of additive helps food stay fresh for longer?**

8 **What method can be used to check for artificial colours in foods?**

9 **Why does biodiesel not put additional carbon dioxide into the atmosphere?**

10 **Why are spillages of biodiesel less harmful to animals than diesel from crude oil?**

1. What are the main layers of the Earth?

2. How thick is the Earth's crust?

3. Which parts of the Earth make up tectonic plates?

4. Where was new evidence for tectonic plates found in the 1960s?

5. What was the main gas in the Earth's early atmosphere?

6. How was oxygen in the atmosphere produced?

7. What is the main gas in the Earth's atmosphere now?

8. Which unreactive gases make up about 1% of the atmosphere?

9. What processes release carbon dioxide into the atmosphere?

10. What processes remove carbon dioxide from the atmosphere?

---

students' book page 96

C1b 6.1 Structure of the Earth

KEY POINTS

1. The Earth is made up of layers – the crust, the mantle and the core.
2. Scientists thought that mountains and valleys formed by the Earth shrinking.

- The Earth is almost spherical with a radius of about 6400 km. At the surface is a thin, solid crust. The thickness of the crust varies between 5 km and 70 km. It is thinnest under the oceans.
- The mantle is under the crust. It is about 3000 km thick, and so goes almost halfway to the centre of the Earth. The mantle is almost entirely solid, but it can flow very slowly.
- The core is very dense and made of metals, mainly nickel and iron. The outer core is liquid and the inner core is solid. This model of the Earth was built up using evidence from seismic waves from earthquakes.

Scientists once thought that mountains and valleys were formed by the Earth shrinking. They believed that the crust solidified as the Earth cooled down and then the Earth continued to shrink, causing the crust to wrinkle.

Key words: crust, mantle, core

GET IT RIGHT!

The inner core is solid and the outer core is liquid.

CHECK YOURSELF

1. Draw a circle with a radius of 6.4 cm to represent the Earth. Draw another circle to show the mantle and core. Calculate how thick the crust should be on your drawing.

2. What is the difference between the outer core and the inner core?

3. Explain how scientists thought that shrinking caused mountains.

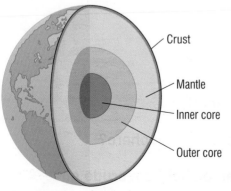

Crust

Mantle

Inner core

Outer core

The structure of the Earth

C1b 6.2 The restless Earth

KEY POINTS

1. The Earth's lithosphere is cracked into pieces called 'tectonic plates'.
2. Convection currents in the mantle cause the tectonic plates to move.
3. Mountains form and earthquakes and volcanoes occur at plate boundaries.
4. Wegener's theory of continental drift was not accepted for many years.

Scientists now believe that mountains form at boundaries between tectonic plates. Tectonic plates are large parts of the lithosphere, the Earth's crust and upper part of the mantle. The lithosphere is broken into tectonic plates that move a few centimetres a year.

Natural radioactivity in the Earth produces heat. This causes convection currents in the mantle that move the plates. At the plate boundaries, huge forces push up the crust to form mountains and cause earthquakes and volcanoes.

Key words: tectonic plates, lithosphere, convection currents, plate boundaries

AQA EXAMINER SAYS…

Alfred Wegener proposed the idea of continental drift in 1915. Other scientists did not accept his ideas. Eventually, in the 1960s, scientists found new evidence deep down on the ocean floor and Wegener's ideas were used to develop the theory of plate tectonics.

CHECK YOURSELF

1. What are tectonic plates?
2. Explain why tectonic plates move.
3. What three things are likely to happen at plate boundaries?
4. Why were Wegener's ideas not accepted for many years?

The distribution of volcanoes around the world largely follows the boundaries of the tectonic plates

C1b 6.3 The Earth's atmosphere in the past

KEY POINTS

1. Volcanoes released most of the gases that formed the Earth's early atmosphere.
2. The gases were mainly carbon dioxide, with some water vapour.
3. The water vapour condensed to form the oceans as the Earth cooled down.
4. Plants evolved and produced oxygen, taking in carbon dioxide.

Scientists think the Earth was formed about 4.5 billion years ago. In the first billion years the surface was covered with volcanoes that released carbon dioxide, water vapour and nitrogen.

As the Earth cooled, most of the water vapour condensed to form the oceans. So the early atmosphere was mainly carbon dioxide with some nitrogen and water vapour. There may have been small amounts of methane and ammonia as well.

In the next two billion years, algae and plants evolved. They used carbon dioxide for photosynthesis to produce food and this released oxygen. As the number of plants increased, the amount of carbon dioxide in the atmosphere decreased and the amount of oxygen increased.

Key words: volcanoes, algae, plants, photosynthesis

AQA EXAMINER SAYS…

There are alternative theories about the Earth's formation and early atmosphere, but many scientists think this one is the most likely.

GET IT RIGHT!

There was very little or no oxygen in the Earth's early atmosphere.

CHECK YOURSELF

1. What produced much of the carbon dioxide, nitrogen and water vapour in the early atmosphere?
2. What happened to most of the water vapour?
3. What process produced oxygen in the atmosphere?

C1b 6.4 Gases in the atmosphere

KEY POINTS

1 The main gases in the atmosphere are nitrogen (78%) and oxygen (21%).
2 There are small amounts of other gases, including carbon dioxide (0.04%), noble gases (almost 1%) and water vapour.

AQA EXAMINER SAYS...

Many students confuse the percentages of oxygen and nitrogen, thinking that there is more oxygen than nitrogen in the air. Also, some think there is much more carbon dioxide than 0.04%.

GET IT RIGHT!

Noble means unreactive.

For the last 200 million years the proportions of the gases in the atmosphere have been about the same as they are now. The atmosphere is almost four-fifths nitrogen and just over one-fifth oxygen.

The other gases make up about 1% of the atmosphere. They are the noble gases (mainly argon), carbon dioxide (0.04%) and water vapour.

Most of the carbon dioxide in the early atmosphere ended up in sedimentary rocks or fossil fuels.

The noble gases are in Group 0 of the periodic table. They are the least reactive elements and they are used where a lack of reactivity is important.

● **Argon** is used as an inert atmosphere, for example in light bulbs to prevent the filament from burning.
● **Neon** is used in electrical discharge tubes for advertising.
● **Helium** is used in balloons.

Key words: noble gases, argon, neon, helium

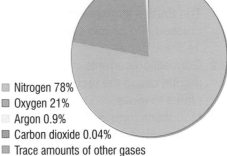

■ Nitrogen 78%
■ Oxygen 21%
■ Argon 0.9%
■ Carbon dioxide 0.04%
■ Trace amounts of other gases

The relative proportions of nitrogen, oxygen and other gases in the Earth's atmosphere

CHECK YOURSELF

1 For how long has the atmosphere been about the same as it is now?
2 What is the third most abundant gas in dry air?
3 Where did most of the carbon dioxide from the early atmosphere end up?
4 Why are the noble gases so useful?

C1b 6.5 The carbon cycle

KEY POINTS

1 The carbon cycle maintains the level of carbon dioxide in the atmosphere.
2 Carbon moves into the atmosphere by respiration, decomposition and combustion.
3 Carbon is removed from the atmosphere by photosynthesis, and by dissolving in water (mainly the oceans).

GET IT RIGHT!

We are releasing the carbon from fossil fuels very quickly compared with the millions of years taken for them to form.

Carbon dioxide moves into and out of the atmosphere and natural processes have kept this in balance for millions of years.

● Plants remove carbon dioxide from the air for photosynthesis to produce food.
● Carbon dioxide is released back into the atmosphere when plants and animals respire or decompose.
● Some of the carbon is used to make animal shells, and in the past these formed sedimentary rocks.
● Volcanoes produce carbon dioxide by decomposing carbonate rocks that have moved deep into the ground.
● Carbon in plants and animals also went into fossil fuels.
● When we burn fossil fuels we release carbon dioxide that had been absorbed from the atmosphere millions of years ago. Combustion of fossil fuels has increased very rapidly in the last 50 years and the level of carbon dioxide in the atmosphere is increasing.
● As the amount of carbon dioxide increases more of it dissolves in the oceans.

Key words: photosynthesis, respire, decompose, combustion, dissolves

The carbon cycle has kept the level of carbon dioxide in the atmosphere steady for the last 200 million years

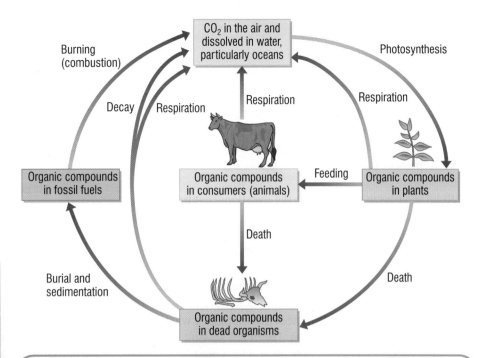

CHECK YOURSELF

1 What happens to most of the carbon dioxide absorbed by plants?

2 What is the main way, other than through plants, that carbon dioxide is removed from the atmosphere?

3 What is the main reason for the recent increase of the amount of carbon dioxide in the atmosphere?

C1b 6   End of chapter questions

1 The average thickness of the Earth's crust is 32 km. What percentage of the Earth's radius is crust?

2 What are the main features of the Earth's mantle?

3 Why do mountains form at tectonic plate boundaries?

4 What caused Wegener's ideas to be accepted in the 1960s?

5 Name three gases that scientists think were in the early atmosphere.

6 Why did the amount of carbon dioxide decrease as plants increased?

7 Nitrogen and oxygen make up most of the atmosphere now. What other gases are in the atmosphere?

8 Why is helium used in balloons?

9 How do volcanoes produce carbon dioxide?

10 How do the oceans remove carbon dioxide from the atmosphere?

1 Draw a circle with a radius of about 4 cm to represent the Earth.

(a) Draw a second circle to represent the core. *(1 mark)*

(b) Label the core, mantle and crust. *(2 marks)*

(c) Why do you not need to draw another circle to represent the crust? *(1 mark)*

2 Read the passage and answer the questions that follow.

Thermoplastics can be melted and reshaped. About 90 % of the energy used in making a thermoplastic food container goes into making the plastic, and about 10% is used to shape the container. These figures are similar for glass, much of which is recycled, but most thermoplastic waste is dumped into landfill sites. There are several reasons why it is more difficult and less cost effective to recycle thermoplastics. Glass can be easily separated into three main types by colour, but it is more difficult to sort thermoplastics because there are many different types and they appear to be very similar. Also, food containers are often made from mixtures of several plastics that can be very difficult to separate. It is more difficult to remove traces of food waste from plastics than from glass. Thermoplastics are lightweight and the average household uses only a small mass of plastic compared with glass.

(a) Why should it be worth recycling thermoplastics? *(1 mark)*

(b) Why is it more difficult to sort thermoplastics than glass? *(2 marks)*

(c) Suggest why it is likely to be less cost effective to recycle plastics than glass. *(2 marks)*

3 The table shows some information about vegetable oils. Use this information to help you to answer the questions.

(a) Which oil is high in polyunsaturated fats but has the lowest yield? *(1 mark)*

(b) Which oil has the most mono-unsaturated fats? *(1 mark)*

(c) Which oil has the most double carbon–carbon bonds and therefore the highest iodine value? *(1 mark)*

(d) Coconut oil has a low enough iodine value to be used in diesel engines without modifying the engine. Why could it not be used in cool climates? *(1 mark)*

(e) How does the melting point of these oils vary with the number of carbon–carbon double bonds they contain? *(1 mark)*

4 Some students made a solution of the colours from a red sweet. They drew a pencil line near the bottom of a piece of chromatography paper. They put a spot of the solution onto the pencil line. They added spots of three common permitted red colours, E123 Amaranth, E124 Ponceau 4R, and E127 Erythrosine. The paper was developed and the results are shown in the diagram.

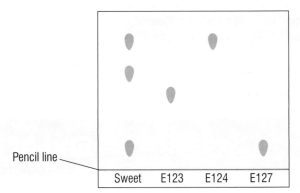

Pencil line

| Sweet | E123 | E124 | E127 |

(a) Describe how the chromatogram is developed. *(3 marks)*

(b) What is the purpose of the pencil line? *(1 mark)*

(c) Why should the line be in pencil and not ink? *(1 mark)*

(d) What colours does the sweet contain? Explain how you can tell from the results. *(3 marks)*

Name of oil	Yield (kg oil per hectare)	Melting point (°C)	Saturated fats (%)	Mono-unsaturated fats (%)	Polyunsaturated fats (%)
coconut	2260	25	85	6	2
corn	145	–15	14	30	51
olive	1019	–6	10	85	0
sunflower	800	–17	8	25	67

Test & Assessment Interactive quizzes, answers and hints online!

The answer is worth 4 marks out of the 5 available. The responses worth a mark are underlined in red.

We can improve the answer in several ways:

Explain, as fully as you can, why cracking is used in the oil industry.

(5 marks)

Cracking is done to change large molecules into small molecules because small molecules have more uses and can be sold at higher prices. Some of the products of cracking are alkanes. Alkanes are saturated hydrocarbons with the formula C_nH_{2n+2} and they can be burnt as fuels. Alkenes are also made. Alkenes are more reactive than alkanes so they can be used to make polymers.

Another possible answer is that **there are more large hydrocarbon molecules produced than can be sold**.

A mark can be gained by adding **alkenes can be used to make other chemicals**.

The formula of alkanes is not relevant to this question and most alkenes can be burnt as fuels. An extra mark can be gained by making it clear that the alkanes produced have **smaller molecules that burn more easily**, so they can be used in high value fuels like petrol.

The answer is worth 4 marks out of the 5 available. The responses worth a mark are underlined in red.

We can improve the answer in several ways:

Describe how to use paper chromatography to find the number of colours in a blue food colouring solution.

(5 marks)

First of all get the chromatography paper and cut it to the right size to fit your beaker. Draw a pencil line near the bottom and put a spot of the blue food colour on the paper on the line. Put a little bit of solvent in the beaker and put the paper in so it just dips into the solvent. Leave it for a few minutes and then check to see how many spots there are. Count the spots to find the number of colours.

This would be better as **cut the paper so that it can stand upright in a beaker**.

Another possible mark would be for writing **cover the beaker with a lid**. There are seven possible marking points for this question.

Instead of leaving it a few minutes, it should read **leave until the solvent reaches the top of the paper**.

There is no need for a pencil line when just checking for colours.

C2 | Additional chemistry (Chapters 1–3)

Checklist

This spider diagram shows the topics in the first half of the unit. You can copy it out and add your notes and questions around it, or cross off each section when you feel confident you know it for your exams.

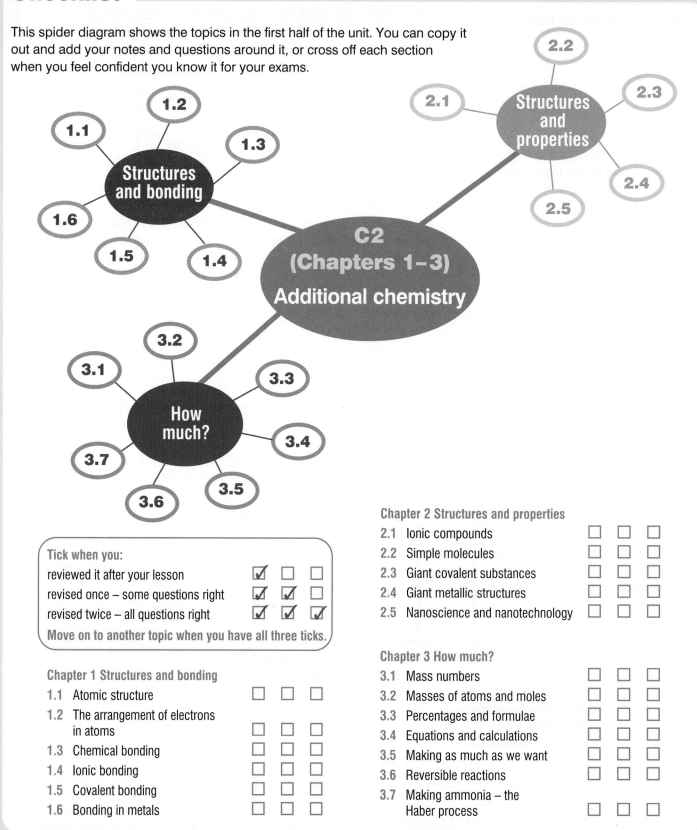

Tick when you:

reviewed it after your lesson	☑	☐	☐
revised once – some questions right	☑	☑	☐
revised twice – all questions right	☑	☑	☑

Move on to another topic when you have all three ticks.

Chapter 1 Structures and bonding

1.1	Atomic structure	☐	☐	☐
1.2	The arrangement of electrons in atoms	☐	☐	☐
1.3	Chemical bonding	☐	☐	☐
1.4	Ionic bonding	☐	☐	☐
1.5	Covalent bonding	☐	☐	☐
1.6	Bonding in metals	☐	☐	☐

Chapter 2 Structures and properties

2.1	Ionic compounds	☐	☐	☐
2.2	Simple molecules	☐	☐	☐
2.3	Giant covalent substances	☐	☐	☐
2.4	Giant metallic structures	☐	☐	☐
2.5	Nanoscience and nanotechnology	☐	☐	☐

Chapter 3 How much?

3.1	Mass numbers	☐	☐	☐
3.2	Masses of atoms and moles	☐	☐	☐
3.3	Percentages and formulae	☐	☐	☐
3.4	Equations and calculations	☐	☐	☐
3.5	Making as much as we want	☐	☐	☐
3.6	Reversible reactions	☐	☐	☐
3.7	Making ammonia – the Haber process	☐	☐	☐

What are you expected to know?

Chapter 1 Structures and bonding (See students' book pages 114–127)

- Atoms have a tiny central nucleus made of protons with a positive charge and neutrons with no charge. The nucleus is surrounded by electrons that have a negative charge, and are equal in number to the protons.
- All atoms of an element have the same number of protons. This is the atomic number (proton number) of the element, and is the order in which elements are arranged in the modern periodic table.
- Electrons are arranged in energy levels (shells). The pattern can be represented by numbers, e.g. 2,8,1 for sodium, or by dot and cross diagrams.
- Elements in the same group in the periodic table have the same number of electrons in their highest energy level (outer shell) and so they have similar chemical properties.
- Ions are formed when atoms lose or gain electrons. Ionic compounds are held together by strong attractions between oppositely charged ions in a giant structure.
- Covalent bonds are formed when pairs of electrons are shared between atoms, and these substances form molecules.
- Metals have giant structures of atoms.
- The atoms (or positively charged ions) are held together by delocalised electrons. This allows metals to conduct heat and electricity. [Higher Tier only]
- The layers of atoms in metals can slide over each other, allowing them to be bent and shaped.

Chapter 2 Structures and properties (See students' book pages 130–139)

- Ionic substances have high melting and boiling points. They conduct electricity when molten or in solution, but not when they are solids.
- Substances made of simple molecules have low melting and boiling points.
- This is because they have weak intermolecular forces. [Higher Tier only]
- The molecules have no charges and so do not conduct electricity.
- Atoms, such as carbon and silicon, that form several covalent bonds can form giant covalent structures.
- Nanoscience involves very small particles made of only a few hundred atoms, which gives these materials special properties and new uses.

Chapter 3 How much? (See students' book pages 142–157)

- Protons and neutrons have a mass of one unit, but electrons have very little mass. The mass number of an atom is the number of protons plus neutrons in the atom.
- Isotopes are atoms of the same element with different mass numbers.
- The relative atomic mass (A_r) of an element is the average mass of its isotopes compared with an atom of the ^{12}C isotope. [Higher Tier only]
- The relative formula mass of a compound (M_r) is found by adding up the relative atomic masses of the atoms shown in its formula.
- One mole of a substance is its A_r or M_r weighed out in grams.
- Percentages of elements in compounds can be calculated from A_r and M_r.
- Percentages by mass can also be used to calculate empirical formulae. [Higher Tier only]
- Masses of reactants and products can be calculated from balanced equations. [Higher Tier only]
- The percentage yield of a reaction does not always equal the theoretical yield. [Higher Tier only]
- The atom economy of a reaction is the mass of the atoms in the useful products as a percentage of the mass of all the atoms in the reactants. [Higher Tier only]
- Reversible reactions can be used efficiently in industrial processes like the Haber process, in which nitrogen and hydrogen react to make ammonia.
- In a closed system a reversible reaction can reach equilibrium. [Higher Tier only]

1. What are the particles in an atom and in which part of an atom are they found?

2. What is the atomic number of an element?

3. How are electrons arranged in atoms?

4. What is special about the electron arrangement of elements in the same group of the periodic table?

5. What is special about the electron arrangement of the noble gases?

6. What happens to electrons when atoms of elements react?

7. What are ionic bonds?

8. Why does sodium chloride have the formula NaCl?

9. What is a covalent bond?

10. Draw a diagram to show the covalent bonds in a molecule of water.

11. How are atoms arranged in metals?

12. How are the atoms in metals held in position within their giant structures? [Higher Tier only]

---

students' book page 114

C2 1.1   Atomic structure

KEY POINTS

1 The nucleus of an atom is made of protons and neutrons.
2 Protons have a positive charge, electrons a negative charge and neutrons are not charged.
3 The atomic number (or proton number) of an element is the number of protons in the nucleus of its atoms.
4 Elements are arranged in order of their atomic numbers in the periodic table.

The nucleus at the centre of an atom contains two types of particle, called protons and neutrons. Protons have a positive charge and neutrons have no charge. Electrons are negatively charged particles that move around the nucleus. An atom has no overall charge, because the number of electrons is equal to the number of protons and their charges are equal and opposite.

All atoms of an element contain the same number of protons. This number is called the atomic number (or proton number) of the element. Elements are arranged in order of their atomic numbers in the periodic table. The atomic number tells you the number of protons and the number of electrons in atoms of the element.

Key words: protons, neutrons, electrons, atomic number, proton number, periodic table

 EXAMINER SAYS...

You can find the atomic number of an element in the periodic table and it tells you the number of protons and the number of electrons in atoms of the element.

CHECK YOURSELF

1 Name the three types of particle in atoms.

2 What are the charges on the three particles?

3 How many electrons are there in an atom of magnesium?

The arrangement of electrons in atoms

KEY POINTS

1 Electrons in atoms are in energy levels that can be represented by shells.
2 Electrons in the lowest energy level are in the shell closest to the nucleus.
3 Electrons occupy energy levels from the lowest first.
4 All the elements in a group of the periodic table have the same number of electrons in their highest energy level (outer shell).

GET IT RIGHT!

Only the first two noble gases have completely full outer shells, but the next energy level begins to fill after each noble gas.

Each electron in an atom is in an energy level. Energy levels can be represented as shells, with electrons in the lowest energy level closest to the nucleus. We can draw them as circles on a diagram, with electrons represented by dots or crosses.

The lowest energy level or first shell can hold two electrons, and the second energy level can hold eight. Electrons occupy the lowest possible energy levels, so the electronic structure of neon with 10 electrons is 2,8. Sodium with 11 electrons has an electronic structure of 2,8,1.

Elements in the same group of the periodic table have the same number of electrons in their highest energy level. All the elements in Group 1 have one electron in their highest energy level, showing that after each noble gas the next energy level begins to fill.

Key words: energy level, shell, electronic structure

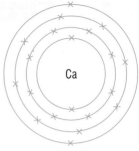

2,8,8,2
Calcium Ca

Once you know the pattern, you should be able to draw the energy levels of the electrons in any of the first 20 atoms (given their atomic number)

CHECK YOURSELF

1 What are electron shells?
2 Draw a diagram to show the electron arrangement of carbon.
3 Write the electronic structures of lithium, nitrogen and magnesium in number form.

Chemical bonding

KEY POINTS

1 Compounds are substances in which elements are chemically combined.
2 When elements react their atoms achieve stable arrangements of electrons.
3 Atoms gain or lose electrons to form ions or share electrons to form covalent bonds.

The noble gases are unreactive because their atoms have stable arrangements of electrons. Atoms of other elements can achieve stable electronic structures by gaining or losing electrons to form ions, or by sharing electrons to form covalent bonds. When an element in Group 1 reacts with an element in Group 7, an electron is transferred between atoms to form ions with the electronic structures of noble gases.

The atoms of elements in Group 1 lose their single outer electron, for example sodium Na (2,8,1) forms sodium ions, Na^+ (2,8). The atoms of elements in Group 7 gain one electron to form ions, for example chlorine Cl (2,8,7) forms chloride ions, Cl^- (2,8,8). We can show this transferring of electrons using dot and cross diagrams. Positive and negative ions attract and form ionic bonds.

Key words: ions, covalent bonds, ionic bonds, sharing, transferring

AQA EXAMINER SAYS...

Ionic bonds hold compounds made of ions together. Other compounds are held together by covalent bonds.

CHECK YOURSELF

1 Why are the noble gases unreactive?
2 Write the formula and electron arrangement for each of: a potassium ion, a magnesium ion, and an oxide ion.
3 Draw a dot and cross diagram to show the formation of lithium fluoride.

C2 1.4 Ionic bonding

KEY POINTS

1 Compounds made of ions have giant structures that are very regular.
2 Strong forces of attraction act throughout the lattice to hold the ions together.

AQA/ EXAMINER SAYS...

Many students seem to think that ionic compounds are made of pairs of ions. This is incorrect because ionic compounds have giant structures. The formula of an ionic compound is the simplest ratio of ions in the compound, and does not represent a molecule.

Ionic bonding holds oppositely charged ions together in giant structures. Strong electrostatic forces of attraction act in all directions. Each ion in the lattice is surrounded by ions with the opposite charge and so is held firmly in place.

The sodium chloride structure contains equal numbers of sodium ions and chloride ions as shown by its formula, NaCl. The sodium ions and chloride ions alternate to form a cubic lattice.

The ratio of ions in the formula and the structure of an ionic compound depend on the charges on the ions. For example, magnesium ions are Mg^{2+}, and chloride ions are Cl^- so the formula of magnesium chloride is $MgCl_2$. Its structure contains twice as many chloride ions as magnesium ions.

Key words: giant structure, lattice, formula

CHECK YOURSELF

1 What forces hold ions together in ionic bonding?

2 What does the formula NaCl tell you about sodium chloride?

3 What is the formula of calcium fluoride?

C2 1.5 Covalent bonding

KEY POINTS

1 A covalent bond is formed when two atoms share a pair of electrons.
2 The number of covalent bonds an atom forms depends on the number of electrons it needs to achieve a stable electron arrangement.
3 Many covalently bonded substances consist of small molecules, but some have giant structures.

The atoms of non-metals need to gain electrons to achieve stable arrangements of electrons. They can do this by sharing electrons with other atoms. Each shared pair of electrons strongly attracts the two atoms, forming a 'covalent bond'.

Atoms of elements in Group 7 need to gain one electron and so form a single covalent bond. Those in Group 6 need to gain two electrons and form two covalent bonds. Atoms of elements in Group 5 can form three bonds and those in Group 4 can form four bonds.

Water
H₂O
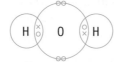

By choosing how we represent a covalent compound, we can show the outer energy level, the shared electrons or just the fact that there are a certain number of covalent bonds

Covalent bonds act only between the two atoms they bond, and so many covalently bonded substances consist of small molecules. Some atoms that can form several bonds, like carbon, can join together in giant covalent structures.

Key words: sharing, covalent bond, molecules, giant covalent structures

AQA/ EXAMINER SAYS...

Covalent bonds join atoms together to form molecules. You should only use the word molecule when describing covalently bonded substances.

CHECK YOURSELF

1 What is a covalent bond?

2 How many covalent bonds can a silicon atom form?

3 Draw a dot and cross diagram for a molecule of ammonia.

C2 1.6 Bonding in metals

KEY POINTS

1 Metals have giant structures of layers of atoms arranged in a regular pattern.
2 The electrons in the highest energy level delocalise. This results in strong electrostatic forces between these electrons and the positively charged metal ions, holding the metal together. [Higher Tier only]

The atoms in a metallic element are all the same size. They form giant structures in which layers of atoms are arranged in regular patterns. Although you cannot see individual atoms, you can see metal crystals on the surfaces of some metals. You can also grow metal crystals by displacement reactions. You can make models of metal structures by putting lots of small spheres like marbles together.

When metal atoms pack together the electrons in the highest energy level (the outer electrons) delocalise and move from one atom to another. This produces positive ions in a 'sea' of moving electrons. The delocalised electrons strongly attract the positive ions and hold the structure together.

Key words: crystals, delocalise

 EXAMINER SAYS...

Some students do not make it clear in their answers that metallic bonding is strong and involves electrostatic forces.

CHECK YOURSELF

1 How are the atoms arranged in metal crystals?
2 Where can you see metal crystals?
3 What are delocalised electrons? [Higher Tier only]
4 What forces hold metal atoms in place in their giant structures? [Higher Tier only]

C2 1 End of chapter questions

1 Why are the numbers of protons and electrons equal in an atom?

2 How many protons and electrons are in an atom of fluorine?

3 What is the arrangement of electrons in an atom of potassium?

4 What is special about the arrangement of electrons of the elements in Group 1?

5 What do we mean by a 'noble gas'?

6 Explain what happens when a sodium atom reacts with a fluorine atom.

7 What holds the ions together in an ionic lattice?

8 Potassium chloride is an ionic compound with formula KCl. What does its formula tell you about the structure of potassium chloride?

9 How is the number of covalent bonds that an atom can form related to its group in the periodic table?

10 Draw a diagram to show the bonds in hydrogen sulphide, H_2S.

11 How are the atoms arranged in a metal crystal?

12 How are the atoms held in position in a metal's giant structure? [Higher Tier only]

1. Why are ionic compounds always solid at room temperature?

2. When can ionic compounds conduct electricity?

3. Why are many covalent substances gases or liquids at room temperature? [Higher Tier only]

4. Why do covalent compounds not conduct electricity?

5. Why can some covalently bonded substances form giant structures?

6. Why are diamond and graphite so different?

7. What happens to the atoms when a metal bends?

8. Do all metals conduct heat and electricity?

9. What is nanoscience?

10. What use can we make of nanoparticles?

students' book
page 130

C2 2.1 Ionic compounds

KEY POINTS

1 Ionic compounds have high melting and boiling points.
2 Ionic compounds conduct electricity when molten or in solution.

AQA EXAMINER SAYS...

There are many strong electrostatic forces of attraction to overcome to melt an ionic solid.

GET IT RIGHT!

Solid ionic compounds cannot conduct electricity.

Ionic compounds have giant structures in which many strong electrostatic forces hold the ions strongly together. This means they are solids at room temperature. A lot of energy is needed to overcome the ionic bonds to melt the solids and so ionic compounds have high melting points and high boiling points.

However, when they have been melted the ions are free to move. This allows them to carry electrical charge, so the liquids conduct electricity. Some ionic solids dissolve in water because water molecules can split up the lattice. The ions are free to move in the solutions and so they also conduct electricity.

Key words: giant structures, ionic bonds, conduct, dissolve

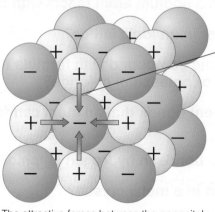

Strong ionic bonds

The attractive forces between the oppositely charged ions in an ionic compound are very strong

CHECK YOURSELF

1 Why do ionic solids have high melting points?

2 Why can some ionic solids dissolve in water?

3 Why can molten ionic substances conduct electricity?

C2 2.2 Simple molecules

The atoms in molecules are held together by strong covalent bonds. These bonds act only between the atoms within the molecule, and so simple molecules have little attraction for each other. Substances made of simple molecules have relatively low melting points and boiling points.

The forces of attraction between molecules, called 'intermolecular forces', are weak.

These forces are overcome when a molecular substance melts or boils. This means that substances made of small molecules have low melting and boiling points. Those with the smallest molecules, like H_2, Cl_2 and CH_4, have the weakest intermolecular forces and are gases at room temperature.

Larger molecules have stronger attractions and so may be liquids at room temperature, like Br_2 and C_6H_{14}, or solids with low melting points, like I_2.

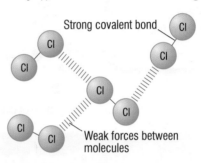

Strong covalent bond

Weak forces between molecules

There are two types of attraction in molecular substances

Key words: intermolecular forces

Molecules have no overall charge and cannot carry an electric current, so these substances do not conduct electricity.

KEY POINTS

1 The forces between simple molecules are weak so many of these substances are gases or liquids at room temperature.
2 Simple molecules do not have a charge and so cannot conduct electricity.

 EXAMINER SAYS...

Many students, when trying to explain melting or boiling, refer to bonds breaking. This suggests they think the covalent bonds are breaking when molecular substances melt, which is not correct. You should make it clear in your answers which forces you are writing about – covalent bonds or intermolecular forces.

GET IT RIGHT!

Covalent bonds are strong and difficult to break; intermolecular forces are much weaker and more easily overcome.

CHECK YOURSELF

1 Why is oxygen, O_2, a gas at room temperature?

2 Why does petrol not conduct electricity?

3 What type of forces act between molecules? [Higher Tier only]

C2 2.3 Giant covalent substances

KEY POINTS

1 Some covalently bonded substances form giant structures.
2 These substances have very high melting points.
3 Diamond and graphite are both forms of carbon but have many different properties.

Atoms of some elements can form several covalent bonds. These atoms can join together in giant covalent structures (sometimes called 'macromolecules'). Every atom in the structure is joined to several other atoms by strong covalent bonds. It takes an enormous amount of energy to break down the lattice and so these substances have very high melting points.

Diamond (a form of carbon) and silica (silicon dioxide) have regular three-dimensional giant structures and so they are very hard and transparent.

Giant covalent substances (cont.)

GET IT RIGHT!

In graphite the atoms are covalently bonded to form flat giant molecules.

Graphite is a form of carbon in which the atoms join in flat two-dimensional layers. There are only weak forces between the layers and so they slide over each other, making graphite slippery and grey.

Key words: macromolecules

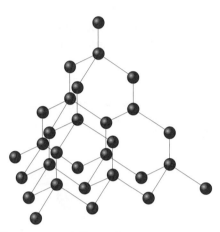

The structure of diamond

Graphite has delocalised electrons (as in a metal structure) along its layers and so conducts electricity.

HIGHER

CHECK YOURSELF

1 What is a macromolecule?

2 Why are diamond and silica transparent and hard?

3 Why is graphite slippery?

4 Why can graphite conduct electricity? [Higher Tier only]

Giant metallic structures

KEY POINTS

1 Metals can be bent and shaped because their layers of atoms can slide over each other.

2 Delocalised electrons move throughout metals and can carry heat and electricity. [Higher Tier only]

Metal atoms are arranged in layers. When a force is applied the layers of atoms can slide over each other. They can move into a new position without breaking apart, so the metal bends or stretches into a new shape. This means that metals are useful for making wires, rods and sheet materials.

 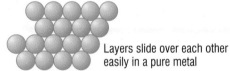

Force — Atoms are all the same size — Pure metal

Layers slide over each other easily in a pure metal

When a force is applied to a metal, the layers slide over each other

Delocalised electrons hold the atoms in place. The delocalised electrons are free to move throughout the metal structure. This means that they can flow as an electric current without changing the metal. They can carry heat energy and so metals are also very good conductors of heat. Many uses of metals depend on their ability to conduct heat and electricity.

HIGHER

Key words: delocalised electrons

CHECK YOURSELF

1 Why can we change the shape of metals?

2 How do metals conduct electricity? [Higher Tier only]

3 What allows metals to conduct heat? [Higher Tier only]

C2 2.5 Nanoscience and nanotechnology

students' book
page 138

KEY POINTS

1 Nanoscience is about structures that are a few nanometres in size.
2 Nanoparticles behave differently to the same materials in bulk.

GET IT RIGHT!

'Nanometre' means one billionth of a metre, i.e. 10^{-9} m.

 EXAMINER SAYS...

You need to know the types of application of nanoparticles, but you do not have to remember details of specific examples. Questions on this topic will often require you to apply your understanding to information supplied in the question.

When atoms are arranged into very small particles they behave differently to ordinary materials made of the same atoms. A nanometre is one billionth of a metre (or 10^{-9} m) and nanoparticles are a few nanometres in size. They contain a few hundred atoms arranged in a particular way. Their structures and very small sizes give them new properties that can make them very useful materials.

Nanoparticles have very large surface areas, exposing many more atoms at their surface than normal materials. Electrons can move through them more easily than ordinary materials. They can be very sensitive to light, heat, pH, electricity and magnetism.

Nanotechnology uses nanoparticles as very selective sensors, highly efficient catalysts, new coatings and construction materials with special properties, and to make drugs more effective.

Key words: nanometre, nanoparticles, nanotechnology

CHECK YOURSELF

1 About how many atoms are there in a typical nanoparticle?
2 Why do nanoparticles have different properties to ordinary materials?
3 Suggest three ways in which nanotechnology is being used.

C2 2 End of chapter questions

1 **Why does it take a lot of energy to melt ionic compounds?**

2 **Why can solutions of ionic compounds conduct electricity?**

3 **What are 'intermolecular forces'? [Higher Tier only]**

4 **What happens to the molecules when water boils? [Higher Tier only]**

5 **Why does diamond have a very high melting point?**

6 **How are the carbon atoms bonded in graphite?**

7 **Why can we pull metals into wires?**

8 **Why do metals stay the same when they conduct electricity? [Higher Tier only]**

9 **How are nanoparticles different to ordinary materials?**

10 **What is 'nanotechnology'?**

1. What is the mass number of an atom?
2. What do we call atoms of the same element with different numbers of neutrons?
3. Why do we use relative atomic masses?
4. What is a mole?
5. How can we find the percentage of an element in a compound?
6. What is the difference between an empirical formula and a molecular formula? [Higher Tier only]
7. What do balanced equations tell us?
8. How much calcium oxide can we make from 10 g of calcium carbonate? [Higher Tier only]
9. What is meant by the yield of a reaction?
10. What do we mean by 'atom economy'?
11. How can we recognise a reversible reaction?
12. What is equilibrium? [Higher Tier only]
13. How is ammonia manufactured?
14. Why is the yield of ammonia so low? [Higher Tier only]

---

students' book page 142

C2 3.1   Mass numbers

KEY POINTS

1. The mass of a proton is equal to the mass of a neutron.
2. The mass number of an atom is the total number of protons and neutrons in its nucleus.
3. Isotopes are atoms of the same element with different numbers of neutrons.

GET IT RIGHT!

number of neutrons = mass number − atomic number

Protons and neutrons have equal masses. The relative masses of a proton and a neutron are both one unit. The mass of an electron is very small compared with a proton or neutron, and so the mass of an atom is made up almost entirely of its protons and neutrons. The total number of protons and neutrons in an atom is called its 'mass number'.

Atoms of the same element all have the same atomic number. The number of protons and electrons in an atom must always be the same, but there can be different numbers of neutrons. Atoms of the same element with different numbers of neutrons are called 'isotopes'. The number of neutrons in an atom is equal to its mass number minus its atomic number. We can show the mass number and atomic number of an atom like this:

$$^{23}_{11}\text{Na}$$

The number at the top (always larger, except in ^1_1H) is the mass number.

Key words: mass number, isotopes

EXAMINER SAYS...

Isotopes are atoms of the same element and have the same chemical properties. They have different physical properties because the different numbers of neutrons give them different masses. Some isotopes are also unstable and radioactive.

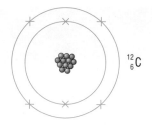

$^{12}_{6}C$

● Proton — Number of protons gives atomic number

● Neutron — Number of protons plus number of neutrons gives mass number

students' book page 144

C2 3.2 Masses of atoms and moles

KEY POINTS

1 Relative atomic masses compare the masses of atoms.
2 The relative atomic mass of an element is an average value for the isotopes of an element. [Higher Tier only]
3 One mole of a substance is its relative formula mass in grams.

AQA EXAMINER SAYS...

Moles are really useful because they give you a way of counting huge numbers of atoms, molecules and ions by weighing the substance in grams.

GET IT RIGHT!

One mole of a substance is its relative formula mass in grams.

EXAM HINTS

● Use the periodic table to look up relative atomic masses.
● When calculating relative formula masses, make sure you add up all the atomic masses in a formula correctly: H_2SO_4 has two hydrogen atoms, one sulfur atom and four oxygen atoms (A_r of H = 1, A_r of O = 16, A_r of S = 32), so its M_r is 98.

Atoms are much too small to weigh and so we use 'relative' atomic masses. These are often shown in periodic tables.

We use an atom of $^{12}_{6}C$ as a standard atom and compare the masses of all other atoms with this.

The relative atomic mass of an element (A_r) is an average value that depends on the isotopes the element contains. However, when rounded to a whole number it is often the same as the mass number of the main isotope of the element.

The relative formula mass (M_r) of a substance is found by adding up the relative atomic masses of the atoms in its formula. For example:

Worked example	Solution
Calculate the M_r of $CaCl_2$	A_r of Ca = 40, A_r of Cl = 35.5 so 40 + (35.5×2) = **111**

The relative formula mass of a substance in grams is called 'one mole' of that substance. Using moles of substances allows us to calculate and weigh out in grams masses of substances with the same number of particles. One mole of sodium atoms contains the same number of atoms as one mole of chlorine atoms. For example:

Worked example	Solution
What is the mass of one mole of NaOH?	A_r of Na = 23, A_r of O = 16, A_r of H = 1 so 23g + 16g + g + 1g = **40g**

Key words: relative atomic mass (A_r), relative formula mass (M_r), mole

CHECK YOURSELF

1 Calculate the relative formula masses (M_r) of:
   (a) H_2 (b) CH_4 (c) $MgCl_2$
   (A_r of H = 1, A_r of C = 12, A_r of Mg = 24, A_r of Cl = 35.5)

2 What is the mass of one mole of water?

3 Atoms of which isotope are used as the standard atoms for relative atomic masses? [Higher Tier only]

CHECK YOURSELF

1 What name do we use for the number of protons and neutrons in an atom?

2 What are isotopes?

3 Calculate the number of protons, neutrons and electrons in these atoms: $^{16}_{8}O$, $^{19}_{9}F$

C2 3.3 Percentages and formulae

KEY POINTS

1 The percentage mass of an element in a compound can be calculated from its A_r and the M_r of the compound.

2 The empirical formula of a compound can be calculated from its percentage composition. [Higher Tier only]

BUMP UP YOUR GRADE

Grade C students should be able to calculate the percentage of an element in a compound. For top grades you should be able to calculate the empirical formula of a compound with two or three elements.

AQA EXAMINER SAYS...

Only higher attaining students will be expected to calculate empirical formulae.

We can calculate the percentage of any of the elements in a compound from the formula of the compound. Divide the relative atomic mass of the element by the relative formula mass of the compound and multiply the answer by 100 to convert it to a percentage. This can be useful when deciding if a compound is suitable for a particular purpose or to identify a compound.

For example:

Worked example

Find the percentage of carbon in carbon dioxide (A_r of C = 12, A_r of O = 16)

Solution

$$M_r \text{ of } CO_2 = 12 + (16 \times 2) = 44$$

$$\text{So percentage of carbon} = \left(\frac{12}{44}\right) \times 100 = \textbf{27.3\%}$$

The empirical formula is the simplest ratio of the atoms or ions in the compound. It is the formula used for ionic compounds, but for covalent compounds it is not always the same as the molecular formula. For example, the molecular formula of ethane is C_2H_6, but its empirical formula is CH_3.

We can calculate the empirical formula of a compound from its percentage composition:

- Divide the mass of each element in 100 g of the compound by its A_r to give the ratio of atoms.
- Then convert this to the simplest whole number ratio.

For example:

Worked example

Work out the empirical formula of the hydrocarbon that contains 80% carbon.

Solution

100 g of hydrocarbon contains 80 g of C and 20 g of H.

$$\text{Number of moles of carbon} = \frac{80}{12} = 6.67$$

$$\text{Number of moles of hydrogen} = \frac{20}{1} = 20$$

Ratio of atoms is 6.67 C : 20 H
Simplest ratio is 1 C : 3 H
So empirical formula is CH_3.

HIGHER

Key words: empirical formula, molecular formula

CHECK YOURSELF

1 What is the percentage by mass of calcium in calcium oxide, CaO?

2 What is the empirical formula of propene, C_3H_6? [Higher Tier only]

3 A compound of iron and chlorine contains 44% iron by mass. What is its empirical formula? [Higher Tier only]

C2 3.4 | **Equations and calculations**

KEY POINTS

1 Balanced chemical equations can be used to calculate masses of reactants and products.
2 In an equation, $2Cl_2$ can mean 2 molecules of chlorine or two moles of chlorine molecules.

 **EXAMINER SAYS...**

You can work in moles or you can use the relative masses in the equation when doing calculations, but don't forget to give correct units in your answer.

EXAM HINTS

Use the periodic table on your Data Sheet (see page 113) to find relative atomic masses if they are not given in the question.

Chemical equations show the reactants and products of a reaction. When they are balanced they show the amounts of atoms, molecules or ions in the reaction.

For example:

$$2Mg + O_2 \rightarrow 2MgO$$

shows that two atoms of magnesium react with one molecule of oxygen to form two magnesium ions and two oxide ions.

Working in relative masses this becomes:

$(2 \times A_r \text{ of Mg}) + (2 \times A_r \text{ of O})$ gives $(2 \times M_r \text{ of MgO})$ or $(2 \times 24 + 2 \times 16 = 2 \times 40)$

If we work in moles, the equation tells us that two moles of magnesium atoms react with one mole of oxygen molecules to produce two moles of magnesium oxide.

This means 48 g of magnesium react with 32 g of oxygen to give 80 g of magnesium oxide. (A_r of Mg = 24, A_r of O = 16)

If we have a known mass of magnesium, say 5 g, we can work out the mass of magnesium oxide using moles.

In this case $5 g = \dfrac{5}{24}$ moles of magnesium and so it will produce:

$$\frac{5}{24} \times 40 g = 8.33 g \text{ of MgO}$$

We can also do it by calculating the proportion of the amounts in the equation:

$$5 g \text{ Mg will produce } 5 \times \frac{80}{48} g = 8.33 g \text{ MgO}$$

CHECK YOURSELF

1 Balance this equation: $H_2 + Cl_2 \rightarrow HCl$

2 Calculate the mass of sodium chloride that can be made from one mole of sodium in the reaction: $2Na + Cl_2 \rightarrow 2NaCl$

3 Calculate the mass of copper oxide that can be made from 10 g of copper in the reaction: $2Cu + O_2 \rightarrow 2CuO$

C2 3.5 | **Making as much as we want**

KEY POINTS

1 The yield of a reaction compares the amount of product actually made with the maximum amount that could be made.
2 Atom economy measures how much of the starting materials becomes useful products.

The yield of a chemical process compares how much you actually make with the maximum amount possible. When you actually carry out chemical reactions it is not possible to collect the amounts calculated from the chemical equations. Reactions may not go to completion and some product may be lost in the process.

The yield is often calculated as a percentage:

$$\text{percentage yield} = \frac{\text{amount of product collected}}{\text{maximum amount of product possible}} \times 100$$

CHECK YOURSELF

1 Why is it not usually possible to collect the maximum yield from a reaction?
2 5.2 g of potassium chloride, KCl, was made from 5.6 g of potassium hydroxide, KOH. What was the percentage yield? [Higher Tier only]
3 Quicklime is produced from limestone by the reaction:
$CaCO_3 \rightarrow CaO + CO_2$
What is the percentage atom economy of this process? [Higher Tier only]

Atom economy

It is also important to consider the amount of the starting materials that ends up in the useful products. This is called the 'atom economy' of a process.

Atom economy is calculated by finding the mass of all the atoms in the starting materials and comparing this with the mass of the atoms in the useful product. It is also often worked out as a percentage:

$$\text{percentage atom economy} = \frac{\text{relative formula mass of useful product}}{\text{relative formula mass of all products}} \times 100$$

Worked example

Zinc is extracted by heating its oxide with carbon. Carbon monoxide is also produced in the reaction.

$$ZnO + C \rightarrow Zn + CO$$

Work out the atom economy from the equation shown above.

(A_r of $Zn = 65$, A_r of $C = 12$, A_r of $O = 16$)

Solution

$$\text{atom economy} = \frac{\text{relative formula mass of useful product (Zn)}}{\text{relative formula mass of all products (Zn + CO)}} \times 100$$

$$= \frac{65}{65 + (12 + 16)} = \frac{65}{93} = \textbf{70\%}$$

● To avoid waste both percentage yield and atom economy should be as high as possible.

Key words: yield, atom economy

HIGHER

Reversible reactions

KEY POINTS

1 Reversible reactions go in both directions.
2 Reversible reactions can reach equilibrium in closed systems. [Higher Tier only]
3 Changing the conditions can change the amounts of reactants and products. [Higher Tier only]

If the products of a chemical reaction can react to produce the reactants the reaction can go in both directions. This type of reaction is called a reversible reaction and is represented with the symbol $\rightleftharpoons$

When there are no products the reaction can only go in the forward direction, but as products build up the reverse reaction can happen. In a closed system nothing can escape and the rates of both forward and backward reactions will become equal. When this happens the system is at equilibrium.

If the conditions of the system are changed the amounts of reactants and products may change. Increasing the concentration of a substance will increase the rate of the reaction away from that substance. If the system is open and products can escape the forward reaction will continue to completion.

CHECK YOURSELF

1 What is a reversible reaction?
2 What can happen to a reversible reaction in a closed system? [Higher Tier only]
3 How can the amounts of reactants and products in a reaction mixture in a closed system be changed? [Higher Tier only]

A reversible reaction

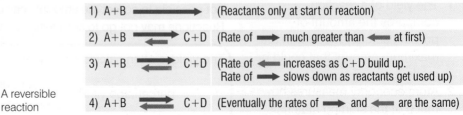

1) A+B ⟶ (Reactants only at start of reaction)

2) A+B ⇌ C+D (Rate of ⟶ much greater than ⟵ at first)

3) A+B ⇌ C+D (Rate of ⟵ increases as C+D build up. Rate of ⟶ slows down as reactants get used up)

4) A+B ⇌ C+D (Eventually the rates of ⟶ and ⟵ are the same)

Key words: reversible, closed system, equilibrium

HIGHER

C2 3.7 Making ammonia – the Haber process

KEY POINTS

1 The Haber process produces ammonia from nitrogen and hydrogen.
2 The reaction uses a high temperature, high pressure and a catalyst to produce a reasonable yield of ammonia in a short time.

AQA EXAMINER SAYS...

Conditions in the Haber process can vary in different places but are always chosen to produce the best yield as quickly as possible.

The Haber process is used to manufacture ammonia, which can be used to make fertilisers and other chemicals.

Nitrogen from the air and hydrogen, which is usually obtained from natural gas, are purified and mixed in the correct proportions. The gases are passed over an iron catalyst at a temperature of about 450°C and a pressure of about 200 atmospheres.

The reaction is reversible and so some ammonia breaks down into nitrogen and hydrogen. The gases are cooled so the ammonia condenses. The liquid ammonia is removed from the unreacted gases and they are recycled.

The yield is less than 20%, but the ammonia is produced quickly and no gases are wasted.

CHECK YOURSELF

1 What are the raw materials used to make ammonia?
2 Write a word equation and a balanced equation for the reaction to make ammonia.
3 Why does the reaction not go to completion?

C2 3 End of chapter questions

1 Why do we not include electrons in mass numbers?

2 How many protons, neutrons and electrons are there in an atom of $^{27}_{13}Al$?

3 What is the relative formula mass of Na_2O?

4 What is the mass of one mole of CO_2?

5 What is the percentage by mass of carbon in methane, CH_4?

6 What is the empirical formula of the compound of iron and oxygen that contains 70% iron? [Higher Tier only]

7 Balance this equation: $CH_4 + O_2 \rightarrow O_2 + H_2O$ [Higher Tier only]

8 Calculate the mass of lithium chloride you can make from 2.4 g of lithium hydroxide: $HCl + LiOH \rightarrow LiCl + H_2O$ [Higher Tier only]

9 Calculate the percentage yield if 4.1 g of zinc was made from 8.1 g of zinc oxide in the reaction: $ZnO + H_2 \rightarrow Zn + H_2O$ [Higher Tier only]

10 What is the percentage atom economy of the reaction to produce copper from copper oxide? $2CuO + C \rightarrow 2Cu + CO_2$ [Higher Tier only]

11 The thermal decomposition of ammonium chloride is a reversible reaction:

$$NH_4Cl \rightleftharpoons NH_3 + HCl$$

Explain what this means.

12 What sort of system is needed for an equilibrium? [Higher Tier only]

13 What conditions are used for the Haber process?

14 What happens to the unreacted gases in the Haber process?

1 Sodium atoms have 11 electrons and oxygen atoms have 8 electrons.

(a) Draw a dot and cross diagram to show the arrangement of electrons in a sodium atom. (2 marks)

(b) Draw a dot and cross diagram to show the arrangement of electrons and the charge for a sodium ion. (2 marks)

(c) Draw a dot and cross diagram to show the arrangement of electrons and the charge on an oxide ion. (2 marks)

(d) Sodium oxide is a solid with a high melting point. Explain why, in terms of its structure and bonding. (2 marks)

(e) Explain why the formula of sodium oxide is Na_2O. (2 marks)

2 (a) Complete the table about the particles in atoms:

Particle	Relative charge	Relative mass
proton	+1	(i)
neutron	(ii)	1
electron	(iii)	very small

(3 marks)

(b) Chlorine has two main isotopes, $^{35}_{17}Cl$ and $^{37}_{17}Cl$.
(i) How many protons, neutrons and electrons are there in an atom of $^{35}_{17}Cl$? (3 marks)
(ii) What is the difference between these two isotopes of chlorine? (1 mark)

(c) Explain why these two isotopes of chlorine have the same chemical properties. (2 marks)

3 Ammonia has the formula NH_3. It is a gas at room temperature.

(a) Name the type of bonding in ammonia. (1 mark)

(b) Draw a dot and cross diagram to represent the bonding in ammonia. (2 marks)

(c) Explain, in terms of its structure, why ammonia is a gas at room temperature. (2 marks)

[Higher]

4 (a) Ammonium chloride, $NH_4Cl(s)$, decomposes when heated to form ammonia gas, $NH_3(g)$ and hydrogen chloride gas $HCl(g)$. When cooled the gases recombine to form ammonium chloride, so the reaction is reversible.

Write a word equation to show this reversible reaction. (2 marks)

(b) Ammonium chloride is used as a fertiliser.
(i) Calculate the mass of one mole of ammonium chloride. (2 marks)
(ii) Calculate the percentage by mass of nitrogen in ammonium chloride. (2 marks)

5 Germanium, Ge, is an element in Group 4 of the periodic table. It is a white, shiny, brittle solid with a very high melting point. It is used in the electronics industry because it conducts a small amount of electricity. It is made from germanium oxide, GeO_2, by reduction with hydrogen. Germanium oxide is an ionic solid and it reacts with hydrochloric acid to produce germanium chloride, $GeCl_4$. Germanium chloride is a volatile liquid. It has small molecules with covalent bonds.

(a) Write a word equation for the reduction of germanium oxide. (2 marks)

(b) Balance the equation for the reaction of germanium oxide with hydrochloric acid:
$$GeO_2 + HCl \rightarrow GeCl_4 + H_2O$$
(2 marks)

(c) Write the formulae of the ions present in germanium oxide. (2 marks)

(d) Draw a dot and cross diagram to show the bonding in germanium chloride. You need to show only the outer electrons. (2 marks)

(e) Germanium has some properties of a metal and some properties of a non-metal. What evidence is there in the information in this question that suggests:
(i) that germanium is a metal? (2 marks)
(ii) that germanium is a non-metal? (2 marks)

(f) Germanium could also be produced by reduction with carbon:
$$GeO_2 + C \rightarrow Ge + CO_2$$
(i) Calculate the percentage atom economy of this reaction. (2 marks)
(ii) Calculate the percentage atom economy for reduction with hydrogen. (2 marks)
(iii) Suggest *two* reasons why reduction with hydrogen is better than reduction with carbon. (2 marks)

[Higher]

Test & Assessment Interactive quizzes, answers and hints online!

The answer is worth 4 marks out of the 5 available. The responses worth a mark are underlined in red.

We can improve the answer in several ways:

Explain, as fully as you can, why sodium chloride has a high melting point.
(5 marks)

Sodium chloride has a giant structure. It is made of sodium ions and chloride ions that have opposite charges. The ions attract each other in the lattice and there are lots of them in a crystal so lots of energy is needed to break them apart and melt them down.

The student has not expressed the ideas sufficiently well to gain all 5 marks. The important point about ionic bonding that has been missed is that the attractions are strong. Adding the word 'strongly' would have scored an additional mark.

This suggests that the student may think that ions are separated when they melt, rather than the forces being overcome sufficiently so that they can move about. Also, the ions themselves do not melt – it is the solid that melts.

The student has scored 4 marks out of a possible 6.

Note that part (c) would only appear on a Higher Tier exam paper.

The responses worth a mark are underlined in red.

We can improve the answer in several ways:

The first part of the working is correct, but the answer is incorrect.

The student uses the answer from part (a) to calculate the percentage of nitrogen. The working is correct using the incorrect answer from (a) and so the error is carried forward and both marks can be awarded. (It is important to carry on working in any calculation using the answers you have written, even if you think your answer to any part may not be correct.)

Ammonium nitrate, NH_4NO_3, is used as a fertiliser.

(a) **Calculate the relative formula mass of ammonium nitrate, NH_4NO_3.**
(2 marks)

(b) **Use your answer to part (a) to calculate the percentage by mass of nitrogen in ammonium nitrate.**
(2 marks)

(c) **A student made 5.2 g of ammonium nitrate crystals by reacting 0.1 mole of nitric acid with ammonia. What was the percentage yield?** *(2 marks)*

(a) $NH_4NO_3 = 14 + 4 + 14 + 16 \times 3$
$= 30 + 48 = 78$

(b) $N + N \quad = 14 + 14 = 28$
$28 \times 100/78 = 35.9\%$

(c) 0.1 mole nitric acid → 0.1 mole ammonium nitrate
0.1 mole $NH_4NO_3 = 7.8\,g$
% yield $= 5.2 \times 100/7.8 = 66\%$

The working is correct using the incorrect answer from (a), but the student has rounded the answer incorrectly. It should be 67% to two significant figures, or 66.7% to three figures.

C2 | Additional chemistry (Chapters 4–7)

Checklist

This spider diagram shows the topics in the second half of the unit. You can copy it out and add your notes and questions around it, or cross off each section when you feel confident you know it for your exams.

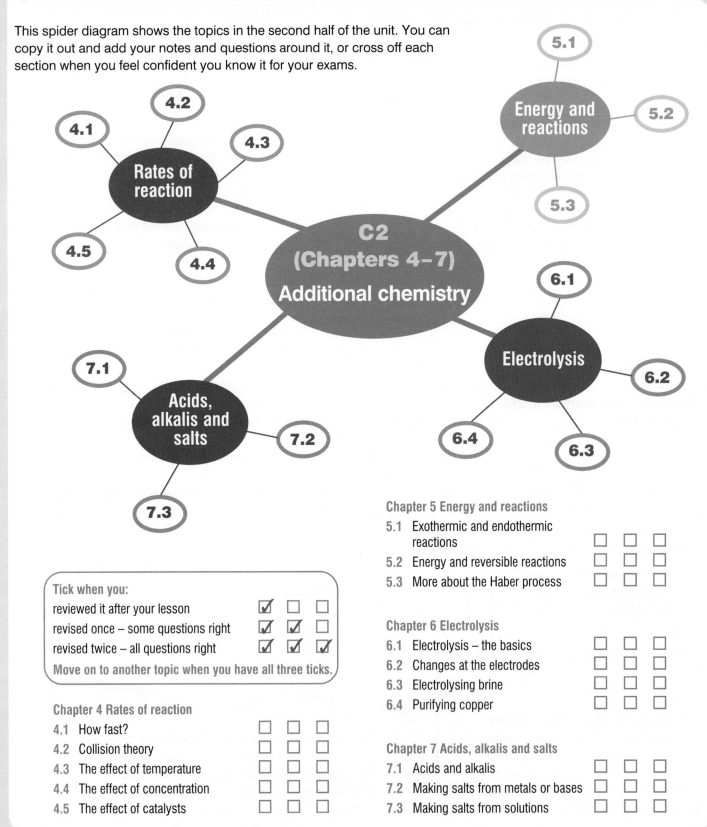

Tick when you:

reviewed it after your lesson	☑	☐	☐
revised once – some questions right	☑	☑	☐
revised twice – all questions right	☑	☑	☑

Move on to another topic when you have all three ticks.

Chapter 4 Rates of reaction

4.1	How fast?	☐	☐	☐
4.2	Collision theory	☐	☐	☐
4.3	The effect of temperature	☐	☐	☐
4.4	The effect of concentration	☐	☐	☐
4.5	The effect of catalysts	☐	☐	☐

Chapter 5 Energy and reactions

5.1	Exothermic and endothermic reactions	☐	☐	☐
5.2	Energy and reversible reactions	☐	☐	☐
5.3	More about the Haber process	☐	☐	☐

Chapter 6 Electrolysis

6.1	Electrolysis – the basics	☐	☐	☐
6.2	Changes at the electrodes	☐	☐	☐
6.3	Electrolysing brine	☐	☐	☐
6.4	Purifying copper	☐	☐	☐

Chapter 7 Acids, alkalis and salts

7.1	Acids and alkalis	☐	☐	☐
7.2	Making salts from metals or bases	☐	☐	☐
7.3	Making salts from solutions	☐	☐	☐

What are you expected to know?

Chapter 4 Rates of reaction (See students' book pages 160–171)

- The rate of a chemical reaction is found by measuring the amount of a reactant used or the amount of a product formed in a unit of time.
- Chemical reactions only happen when particles collide with enough energy to react. The minimum energy they need to react is called the 'activation energy'.
- Increasing temperature, concentration of solutions, pressure of gases, surface area of solids and using a catalyst increases the rate of reactions.
- Increasing the temperature of reactants causes the particles to collide more often and with more energy.
- Increasing the concentration of reactants, pressure of gases, or the surface area of solid reactants causes particles to collide more frequently.
- Concentrations of solutions measured in moles per dm^3 enable you to compare the number of particles of the substances dissolved in the solution. [Higher Tier only]
- Equal volumes of gases at the same temperature and pressure contain the same number of molecules (and the same number of moles). [Higher Tier only]
- Catalysts change the rate of particular reactions and are important in speeding up industrial processes.

Chapter 5 Energy and reactions (See students' book pages 174–181)

- Exothermic reactions transfer heat to the surroundings, and endothermic reactions take in heat from the surroundings.
- Reversible reactions are exothermic in one direction and endothermic in the other direction.
- If the temperature is raised, the yield from the endothermic reaction increases and the yield from the exothermic reaction decreases. [Higher Tier only]
- In reversible reactions involving gases, high pressure favours the reaction that produces the smallest number of molecules of gas. [Higher Tier only]
- It is important to minimise the energy used and the energy wasted in industrial processes for economic and environmental reasons.

Chapter 6 Electrolysis (See students' book pages 184–193)

- Ions in molten ionic compounds or in solutions are free to move and so can conduct electricity. The compounds are broken down into elements by electrolysis.
- Positive ions are attracted to the negative electrode where they gain electrons (reduction) and negative ions lose electrons at the positive electrode (oxidation).
- Reactions at electrodes depend on the ions that are present, their concentrations and what the electrode is made from.
- Electrolysis of sodium chloride solution is used in industry to produce chlorine, sodium hydroxide and hydrogen.
- Copper can be purified by electrolysis using copper electrodes in a solution containing copper ions.
- Reactions at electrodes can be represented by half equations. [Higher Tier only]

Chapter 7 Acids, alkalis and salts (See students' book pages 196–203)

- Insoluble salts can be made by mixing solutions of ions to form a precipitate.
- Soluble salts can be made from acids using a suitable metal, an insoluble base or an alkali. Then the solution can be crystallised to obtain the salt.
- Hydrogen ions $H^+(aq)$ make solutions acidic and hydroxide ions $OH^-(aq)$ make solutions alkaline. The pH scale measures the acidity or alkalinity of a solution.
- Neutralisation is the reaction of $H^+(aq)$ ions with $OH^-(aq)$ ions to produce $H_2O(l)$.

(1) **What is the rate of a reaction?**

(2) **How can we measure the rate of a reaction?**

(3) **What must happen to particles for them to react?**

(4) **Why does changing the conditions change the rate of a reaction?**

(5) **Why does increasing the temperature increase the rate of a reaction?**

(6) **What happens to the rate of many reactions if you increase the temperature by 10°C?**

(7) **What happens to the rate of a reaction if you increase the concentration of the reactants?**

(8) **How does changing the pressure affect the rate of reactions between gases?**

(9) **What is a catalyst?**

(10) **Why are catalysts used in many industrial processes?**

---

students' book
page 160

C2 4.1   How fast?

KEY POINTS

1 The rate of a reaction tells us how quickly reactants become products.
2 We can measure how quickly reactants are used up or how quickly products are formed.
3 Measuring a rate involves measuring an amount and the time it takes.

EXAMINER SAYS...

The faster the rate, the shorter the time it takes for the reaction. So rate is inversely proportional to time.

The rate of a reaction measures the speed of a reaction or how fast it is. The rate can be found by measuring how much of a reactant is used up or how much of a product is formed in a certain time.

An alternative way is to measure the time for a certain amount of reactant to be used or product to be formed.

Whichever way it is done it involves measuring both an amount and a time because:

$$\text{rate} = \frac{\text{amount of reactant used}}{\text{time}} \quad \text{OR} \quad \frac{\text{amount of product formed}}{\text{time}}$$

The simplest measurements we can make are the mass of gas released or the volume of gas produced at intervals of time. Another method is to measure the time it takes for a certain amount of solid to appear in a solution. Other possible ways include measuring changes in the colour, concentration, or pH of a reaction mixture over time.

Key words: rate of reaction

CHECK YOURSELF

1 Explain what we mean by the 'rate of a reaction'.

2 What two things must we measure to find the rate of a reaction?

3 How can we measure the rate of a reaction that gives off a gas?

C2 4.2 Collision theory

KEY POINTS

1 Reactions happen when particles collide with enough energy to bring about a change.
2 Factors that affect the frequency or energy of collisions will change the rate of a reaction.

The collision theory states that reactions can only happen if particles collide. However, just colliding is not enough. The particles must collide with enough energy to change into new substances. The minimum energy they need is called the 'activation energy'.

Factors that increase the chance of collisions or the energy of the particles will increase the rate of the reaction. Increasing the temperature, concentration of solutions, pressure of gases, surface area of solids and using a catalyst will increase the rate of a reaction.

Breaking large pieces of a solid into smaller pieces exposes new surfaces and so increases the surface area. This means there are more collisions in the same time and so a powder reacts faster than large lumps of a substance. There are many examples of powders reacting very rapidly.

Key words: collision theory, activation energy, surface area

GET IT RIGHT!

Particles must collide with more than the minimum energy needed for a reaction to happen.

CHECK YOURSELF

1 What is the collision theory of reactions?
2 What factors increase the rate of a reaction?
3 Why do powders react faster than solids?

C2 4.3 The effect of temperature

KEY POINTS

1 Increasing the temperature increases the rate of reactions.
2 A small increase in temperature produces a large change in reaction rates.

Increasing the temperature increases the speed of the particles in a reaction mixture. This means they collide more often, which increases the rate of reaction. As well as colliding more frequently they collide with more energy, which also increases the rate of reaction.

Therefore, a small change in temperature has a large effect on reaction rates. At ordinary temperatures a rise of 10°C will roughly double the rate of many reactions, so they go twice as fast. A decrease in temperature will slow reactions down, and a change of 10°C will double the time that many reactions take. This is why we refrigerate or freeze food so it stays fresh for longer.

CHECK YOURSELF

1 Why does a small change in temperature have a large effect on reaction rates?
2 What temperature change doubles the rate of many reactions?
3 Explain why refrigerating food makes it last longer.

Cold – slow movement, less frequent collisions, little energy

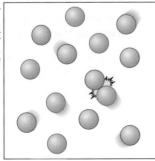
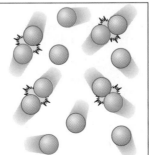

Hot – fast movement, more frequent collisions, more energy

More frequent collisions with more energy – both of these increase the rate of a chemical reaction as the temperature increases

KEY POINTS

1 Increasing the concentration of reactants increases the rate of reactions.
2 Concentrations of solutions are measured in moles per cubic decimetre (mol/dm³). [Higher Tier only]
3 Equal volumes of gases at the same temperature and pressure contain equal numbers of molecules. [Higher Tier only]

AQA EXAMINER SAYS...

You will not have to work out the number of moles of substances in solutions or in gases in this unit. However, it is important that you understand that equal volumes of solutions of equal concentrations contain the same number of particles. You also need to know that if the concentration is doubled the number of particles in a given volume is doubled.

If the concentration of a solution is increased there are more particles dissolved in the same volume. This means the dissolved particles are closer together and so they collide more often. Increasing the concentration of a reactant therefore increases the rate of a reaction.

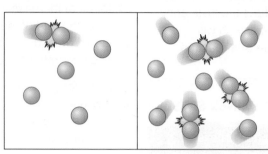

Low concentration/ low pressure — High concentration/ high pressure

Increasing concentration and pressure mean that particles are closer together. This increases the frequency of collisions between particles, so the reaction rate increases.

In a similar way, increasing the pressure of a gas puts more molecules into the same volume, and so they collide more frequently. This increases the rate of reactions that have gases as reactants.

Equal volumes of gases at the same temperature and pressure contain equal numbers of molecules. So 10 cm³ of hydrogen contains the same number of molecules as 10 cm³ of oxygen at the same temperature and pressure.

Measuring concentrations of solutions in moles per cubic decimetre (mol/dm³) means we can measure out equal numbers of particles of the solutes by taking equal volumes of the same concentration.

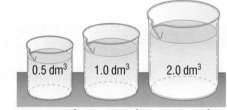

0.5 dm³   1.0 dm³   2.0 dm³

2.0 mol/dm³ 1.0 mol/dm³ 0.5 mol/dm³

These different volumes of solution all contain the same amount of solute – but at different concentrations

The graph below shows how the mass of a reaction mixture changes as a gas is given off. The three lines are drawn for different concentrations of a solution reacting with an excess of solid reactant:

Mass

Lower acid concentration

Higher acid concentration

Time

CHECK YOURSELF

1 Why does increasing the concentration of a reactant increase the rate of the reaction?

2 Why does increasing the pressure of a gas reactant increase the rate of the reaction?

3 Why is it useful to measure concentrations in moles per cubic decimetre? [Higher Tier only]

C2 4.5 The effect of catalysts

KEY POINTS

1 Catalysts change the rates of chemical reactions but are not used up in the reaction.
2 Catalysts that speed up reactions are important in industry to reduce costs.

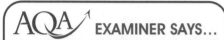

EXAMINER SAYS...

Most of the catalysts that are used are positive catalysts that increase the rates of reactions. (Negative catalysts or inhibitors are used to slow reactions down, but are not studied at GCSE.)

Catalysts change the rates of chemical reactions. Most catalysts are used to speed up reactions. The catalyst is left at the end of the reaction and so it can be used over and over again. Catalysts work by lowering the activation energy of a reaction so that more collisions result in a reaction.

Although some catalysts are expensive, they can be economical because they do not need replacing very often. They can also reduce the energy costs and time needed for a reaction. Catalysts often work with only one type of reaction and so different reactions need different catalysts. Finding new and better catalysts is a major area of research for the chemical industry.

Key words: catalyst, activation energy

GET IT RIGHT!

Catalysts are left at the end of a reaction but they definitely play a part in the reaction.

CHECK YOURSELF

1 What is the effect of a catalyst on the activation energy of a reaction?

2 Why can expensive catalysts be economical?

3 Why are many different catalysts needed?

C2 4 End of chapter questions

1 Write an equation to show how to calculate the rate of a reaction.

2 Suggest one way that you could measure the rate of reaction between magnesium and hydrochloric acid: $Mg + 2HCl \longrightarrow MgCl_2 + H_2$

3 What name is used for the minimum energy needed for a reaction to happen?

4 How could you increase the rate of the reaction between magnesium and hydrochloric acid?

5 Why do many reactions go twice as fast if the temperature is increased by 10°C?

6 What will happen to the rate of a reaction if the temperature is decreased from 30°C to 10°C?

7 Explain why the frequency of collisions increases if you increase the concentration of a solution.

8 Why do gases react faster at higher pressures?

9 How do catalysts speed up reactions?

10 Why do catalysts not need replacing very often?

1. What name do we use for reactions that transfer energy to the surroundings?

2. What type of reaction causes the temperature of the surroundings to decrease?

3. In a reversible reaction, how does the energy of the forward reaction compare with the energy of the reverse reaction?

4. How does increasing the temperature affect a reversible reaction in a closed system? [Higher Tier only]

5. What sort of reversible reactions are affected by pressure? [Higher Tier only]

6. Why are compromise conditions used in the process to make ammonia? [Higher Tier only]

---

students' book page 174

C2 5.1 Exothermic and endothermic reactions

KEY POINTS

1 Exothermic reactions transfer energy to the surroundings.
2 Endothermic reactions transfer energy from the surroundings.

AQA↗ EXAMINER SAYS...

You should know the main types of reaction that are exothermic and endothermic reactions. You should be able to recognise a reaction as exothermic or endothermic from information about temperature changes.

EXAM HINTS

Endothermic reactions take in heat. Heat exits (transfers from) exothermic reactions.

When chemical reactions take place energy is transferred as bonds are broken and made. Reactions that transfer energy to the surroundings are called 'exothermic' reactions. The energy transferred often heats up the surroundings and so the temperature increases. Exothermic reactions include combustion, such as burning fuels and metals, respiration and neutralisation.

Endothermic reactions take in energy from the surroundings. Some cause a decrease in temperature and others require a supply of energy. When some solid compounds are mixed with water, the temperature decreases because endothermic reactions happen. Thermal decomposition reactions need a supply of heat to keep going. Photosynthesis is an important endothermic reaction that uses light energy.

Key words: exothermic, endothermic

When a fuel burns in oxygen, energy is transferred to the surroundings

When we eat sherbet we can feel an endothermic reaction! Sherbet dissolving in the water in your mouth takes in energy – giving a slight cooling effect.

Investigating energy changes

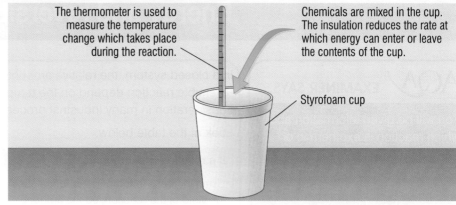

The thermometer is used to measure the temperature change which takes place during the reaction.

Chemicals are mixed in the cup. The insulation reduces the rate at which energy can enter or leave the contents of the cup.

Styrofoam cup

We can use very simple apparatus to investigate the energy changes in reactions. Often we don't need to use anything more complicated than a Styrofoam drinks cup and a thermometer.

CHECK YOURSELF

1 What do we call reactions that transfer energy from the surroundings?

2 How do we know that burning methane is exothermic?

3 Why do some sweets produce a cooling effect in your mouth?

---

students' book page 176 **C2 5.2** # Energy and reversible reactions

KEY POINTS

1 Reversible reactions are exothermic in one direction and endothermic in the other direction.

2 Increasing the temperature increases the amount of products from the endothermic reaction. [Higher Tier only]

GET IT RIGHT!

The energy changes of the forward and reverse reactions are always equal.

In reversible reactions, the forward and reverse reactions involve equal but opposite energy transfers. A reversible reaction that is exothermic in one direction must be endothermic in the other direction. The amount of energy released by the exothermic reaction exactly equals the amount taken in by the endothermic reaction.

Changing the temperature of a reversible reaction in a closed system changes the amounts of the reactants and products. If we increase the temperature, the amount of products from the endothermic reaction increases. If we decrease the temperature, the amount of products from the exothermic reaction increases. This means that we can change the yield of the reaction by changing the temperature.

- Heating blue copper sulfate crystals is an endothermic reaction:

$$\underset{\substack{\text{hydrated}\\\text{copper sulfate}}}{\underset{\text{blue crystals}}{CuSO_4.5H_2O}} \rightleftharpoons \underset{\substack{\text{anhydrous}\\\text{copper sulfate}}}{\underset{\text{white powder}}{CuSO_4}} + 5H_2O$$

- Adding water to anhydrous copper sulfate is an exothermic reaction.

In a closed system, the relative amounts of reactants and products in a reversible reaction depend on the temperature. This is a very important consideration in many industrial processes.

Look at the table below:

If a reaction is exothermic	If a reaction is endothermic ......
... an increase in temperature decreases the yield of the reaction, so the amount of products formed is lower.	... an increase in temperature increases the yield of the reaction, so the amount of products formed is larger.
... a decrease in temperature increases the yield of the reaction, so the amount of products formed is larger.	... a decrease in temperature decreases the yield of the reaction, so the amount of products formed is lower.

CHECK YOURSELF

1 In a reversible reaction the forward reaction is endothermic. What does this tell you about the reverse reaction?

2 In a reversible reaction the forward reaction releases 50 kJ of energy. What will be the energy transfer for the reverse reaction?

3 In the reversible reaction $H_2 + I_2 \rightleftharpoons 2HI$ the formation of HI is exothermic. What should you do to the temperature to increase the yield of HI? [Higher Tier only]

KEY POINTS

1 For reversible reactions involving gases, increasing the pressure increases the yield of the reaction that produces the smaller number of molecules of gas.

2 The conditions for industrial processes are chosen to give as much product as possible as quickly as possible for the lowest cost.

Changes in pressure affect the yield of reversible reactions that have different numbers of molecules of gases in the reactants and products. An increase in pressure will increase the yield of a reaction that has fewer molecules of gases in the products than in the reactants.

If a reaction produces a larger volume of gases	If a reaction produces a smaller volume of gases ......
... an increase in pressure decreases the yield of the reaction, so the amount of products formed is lower.	... an increase in pressure increases the yield of the reaction, so the amount of products formed is larger.
... a decrease in pressure increases the yield of the reaction, so the amount of products formed is larger.	... a decrease in pressure decreases the yield of the reaction, so the amount of products formed is lower.

In the reaction for the Haber process:

$$N_2 + 3H_2 \rightleftharpoons 2NH_3$$

four molecules of reactant gases produce two molecules of ammonia gas. So increasing the pressure will produce more ammonia. However, increasing the pressure increases the costs of the process and so a compromise of a reasonably high pressure is used.

GET IT RIGHT!

Changing the pressure only affects the yield in reactions with different numbers of molecules of gases in the reactants compared with the products.

The reaction to produce ammonia is exothermic, so lower temperatures give higher yields. However, the reaction is slower at lower temperatures because the rate decreases and the catalyst does not work as well, so a compromise temperature is used. The conditions are chosen to produce ammonia as economically as possible. Industrial processes are being developed that use low temperatures and pressures to reduce energy use and waste.

It is expensive to build chemical plants that operate at high pressures

CHECK YOURSELF

1 How can you tell if a reversible reaction will be affected by changes in pressure?

2 Why do many industrial processes making ammonia operate at about 200 atmospheres pressure?

3 Why is a temperature of about 450°C used to make ammonia?

C2 5 — End of chapter questions

1 **Name two types of reaction that are exothermic.**

2 **Why do some solids produce a cooling effect when mixed with water?**

3 **The reaction $2SO_2 + O_2 \rightleftharpoons 2SO_3$ is exothermic in the forward direction. What should be done to the temperature to obtain a high yield of SO_3? [Higher Tier only]**

4 **In the forward reaction in question 3, there are 95 kJ of energy released. What is the energy change for the reverse reaction?**

5 **The reaction $H_2 + I_2 \rightleftharpoons 2HI$ is reversible. How would increasing the pressure affect the yield of HI? [Higher Tier only]**

6 **The reaction $2NO_2 \rightleftharpoons N_2O_4$ is exothermic. What conditions of temperature and pressure would increase the yield of N_2O_4? [Higher Tier only]**

(1) **What happens to ionic compounds when they are electrolysed?**

(2) **What substances can be produced at the positive electrode in electrolysis?**

(3) **Where does reduction take place during electrolysis?**

(4) **What sort of electrolyte can produce hydrogen at the negative electrode?**

(5) **What products are formed when brine is electrolysed?**

(6) **Why is electrolysis of brine done on a large scale in industry?**

(7) **What change takes place at the positive electrode in the purification of copper by electrolysis?**

(8) **What happens to the impurities from the copper?**

---

students' book page 184

C2 6.1   Electrolysis – the basics

KEY POINTS

1 Electrolysis decomposes ionic substances into elements.
2 Ions are free to move when ionic solids are molten or dissolved in water.
3 Metals or hydrogen are formed at the negative electrode and non-metallic elements are formed at the positive electrode.

EXAMINER SAYS...

Electrodes are usually made of substances that will not react with the electrolyte or the elements that are formed.

When electricity is passed through a molten ionic compound or a solution containing ions, electrolysis takes place. The molten ionic solid or solution of ions is called the 'electrolyte'.

The electrical circuit has two conducting rods called 'electrodes' that make contact with the electrolyte. The ions in the electrolyte move to the electrodes where they are discharged to produce elements.

- Positively charged ions are attracted to the negative electrode where they form metals. Hydrogen may be formed at the negative electrode if the ions are dissolved in water.
- Negatively charged ions are attracted to the positive electrode where they lose their charge to form non-metallic elements.
- For example, when molten lead bromide is electrolysed, lead and bromine are produced.

Key words: electrolysis, electrolyte, electrode

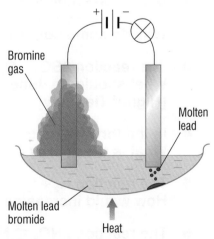

When we pass electricity through molten lead bromide it forms molten lead and brown bromine gas, as the electrolyte is broken down by the electricity

GET IT RIGHT!

Ions are discharged at the electrodes to produce elements.

CHECK YOURSELF

1 What is an electrolyte?

2 What happens to positively charged ions during electrolysis?

3 Which elements are produced when molten zinc chloride is electrolysed?

students' book page 186

C2 6.2 Changes at the electrodes

HIGHER

KEY POINTS

1 Positively charged ions gain electrons at the negative electrode.
2 Negatively charged ions lose electrons at the positive electrode.
3 Gaining electrons is reduction; loss of electrons is oxidation.
4 Electrolysing solutions of ions in water may produce hydrogen.

EXAMINER SAYS...

Only Higher Tier students will be expected to write half equations for reactions at electrodes in the examination.

EXAM HINTS

In half equations the number of electrons must balance the number of charges on the ion.

GET IT RIGHT!

Oxidation Is Loss of electrons, Reduction Is Gain – OIL RIG.

- When positively charged ions reach the negative electrode they gain electrons to become neutral atoms. Gaining electrons is called 'reduction', so the positive ions have been reduced. Ions with a single positive charge gain one electron and those with a 2+ charge gain 2 electrons.
- At the positive electrode, negative ions lose electrons to become neutral atoms. This is 'oxidation'. Some non-metal atoms combine to form molecules, for example bromine forms Br_2.

We can represent the changes at the electrodes by half equations. The equations for lead bromide are:

At the negative electrode: $Pb^{2+} + 2e^- \rightarrow Pb$

At the positive electrode: $2Br^- \rightarrow Br_2 + 2e^-$

Water contains hydrogen ions and hydroxide ions. When solutions of ions in water are electrolysed, hydrogen may be produced at the positive electrode. This happens if the other positive ions are of metals more reactive than hydrogen.

Key words: reduction, oxidation

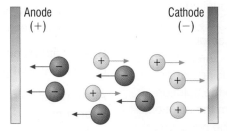

An ion always moves towards the oppositely charged electrode

CHECK YOURSELF

1 What happens in electrolysis to copper ions, Cu^{2+}, at the negative electrode?

2 What happens in electrolysis to chloride ions, Cl^-, at the positive electrode?

3 Why are ions of metals always reduced in electrolysis?

C2 6.3 Electrolysing brine

HIGHER

KEY POINTS

1 Electrolysing brine produces hydrogen, chlorine and sodium hydroxide.
2 The products have many important uses.

EXAMINER SAYS...

You may be asked to apply your knowledge of electrolysis to other industrial processes in the examination.

Brine is a solution of sodium chloride in water. When it is electrolysed, hydrogen is produced at the negative electrode from hydrogen ions in the water. Chlorine is produced at the positive electrode from the chloride ions. This leaves sodium ions and hydroxide ions (from water) in the solution.

The half equations for the reactions at the electrodes are:

At the positive electrode: $2Cl^- \rightarrow Cl_2 + 2e^-$

At the negative electrode: $2H^+ + 2e^- \rightarrow H_2$

● Sodium hydroxide is a strong alkali and has many uses including making soap, making paper, making bleach, neutralising acids and controlling pH.
● Chlorine is used to kill bacteria in drinking water and swimming pools, and to make bleach, disinfectants and plastics.
● Hydrogen is used to make margarine and hydrochloric acid.

Key words: brine, hydrogen, chlorine, sodium hydroxide

CHECK YOURSELF

1 What are the four ions present in brine?

2 Explain why sodium hydroxide is left in the solution.

3 Give two uses for each of the products.

C2 6.4 Purifying copper

HIGHER

KEY POINTS

1 Copper can be purified by electrolysis using a solution containing copper ions.
2 Impure copper is the positive electrode, producing copper ions that go into the solution.
3 Copper ions from the solution are reduced at the negative electrode to form pure copper.

Impurities in copper affect its properties including its conductivity. Copper for use as electrical wires must be very pure. It can be purified by electrolysis, using copper electrodes in a solution of a copper salt. The impure copper is used as the positive electrode and the negative electrode is a thin sheet of pure copper.

Copper atoms on the positive electrode are oxidised, losing electrons to form copper ions that go into the solution.

At the negative electrode copper ions from the solution are reduced, forming copper metal.

The copper is deposited on the negative electrode, which increases in thickness.

The half equations for the reactions at the electrodes are:

At the positive electrode: $Cu(s) \rightarrow Cu^{2+}(aq) + 2e^-$

At the negative electrode: $Cu^{2+}(aq) + 2e^- \rightarrow Cu(s)$

As the copper from the positive electrode dissolves, the impurities are released and collect as sludge at the bottom of the cell. The impurities include precious metals like gold, silver and platinum. These are extracted from the sludge.

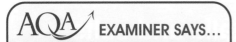
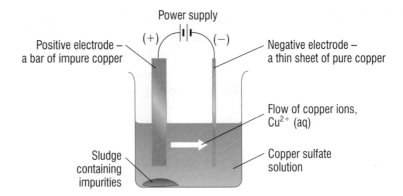

Copper is refined using electrolysis

CHECK YOURSELF

1 Why is it necessary for copper wires to be very pure?

2 Explain why copper ions from the positive electrode go into the solution when copper is purified by electrolysis.

3 Why does the negative electrode increase in thickness?

C2 6	End of chapter questions

1 **What happens to the ions in an electrolyte during electrolysis?**

2 **What is produced at the positive electrode when molten magnesium chloride is electrolysed?**

3 **Explain what happens to magnesium ions at the negative electrode when magnesium chloride is electrolysed.**

4 **What product is formed by oxidation when copper bromide is electrolysed?**

5 **Why does electrolysing brine give different products to electrolysing molten sodium chloride?**

6 **Which products from electrolysing brine are used to make bleach?**

7 **What happens to copper atoms from the impure copper electrode in the purification of copper by electrolysis?**

8 **Why is pure copper deposited on the other electrode?**

9 **Copy and complete the balanced half equation for the deposition of aluminium at the negative electrode:**

$$Al^{3+} \rightarrow Al$$ **[Higher Tier only]**

10 **Copy and complete the balanced half equation that shows the release of chlorine gas at the positive electrode:**

$$Cl^- \rightarrow Cl_2$$ **[Higher Tier only]**

1. Which ions make solutions:

   (a) acidic   (b) alkaline?

2. A solution has a pH of 12. What does this tell you about the solution?

3. What are the products when an acid reacts with a metal?

4. Name a base that could be used to make magnesium chloride.

5. What are the products formed when potassium hydroxide reacts with sulfuric acid?

6. How can you make the insoluble salt lead chloride?

---

students' book
page 196

C2 7.1   Acids and alkalis

KEY POINTS

1. When acids are added to water they form hydrogen ions, $H^+(aq)$.
2. When alkalis are added to water they form hydroxide ions, $OH^-(aq)$.
3. The pH scale measures the acidity or alkalinity of solutions in water.

AQA ✓ EXAMINER SAYS...

A base is a substance that will neutralise an acid, but some bases do not dissolve in water. Only bases that dissolve in water are called alkalis.

GET IT RIGHT!

Acids have pH values below 7. Alkalis have pH values greater than 7.

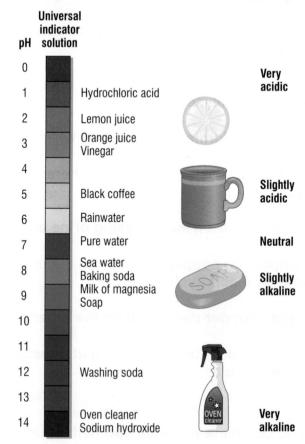

Universal indicator solution

pH		
0		Very acidic
1	Hydrochloric acid	
2	Lemon juice	
3	Orange juice / Vinegar	
4		
5	Black coffee	Slightly acidic
6	Rainwater	
7	Pure water	Neutral
8	Sea water / Baking soda	
9	Milk of magnesia / Soap	Slightly alkaline
10		
11		
12	Washing soda	
13		
14	Oven cleaner / Sodium hydroxide	Very alkaline

The pH scale tells us how acidic or alkaline a solution is

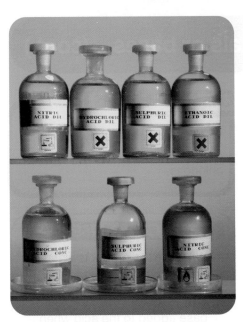

Some common laboratory acids

- Pure water is neutral and has a pH value of 7.
- Acids are substances that produce hydrogen ions, $H^+(aq)$, when they are added to water. This makes the solution acidic and it has a pH value of less than 7.
- Bases react with acids and neutralise them. Alkalis are bases that dissolve in water to make the solution alkaline. They produce hydroxide ions, $OH^-(aq)$, in the solution. Alkaline solutions have a pH value greater than 7.

The pH scale has values from 0 to 14. Solutions that are very acidic have low pH values between 0 and 2, and solutions that are very alkaline have high pH values of 12 to 14. Indicators have different colours in acidic and alkaline solutions. Universal indicator and full-range indicators have different colours at different pH values.

Key words: hydrogen ions, hydroxide ions, pH, neutral, neutralise, indicators

CHECK YOURSELF

1 When a solid dissolves in water the solution has a pH value of 3.5. Name the ions that give this pH value.

2 Why does the pH of water change when alkalis dissolve?

3 Name a substance that can be used to test the pH of a solution.

---

students' book page 198 C2 7.2 **Making salts from metals or bases**

KEY POINTS

1 Salts are formed when acids react with metals or bases.

2 All of the acid can be used up if we add excess solid.

3 The salt that is produced depends on the acid and the metal in the reactants.

AQA EXAMINER SAYS...

You should find it easy to work out the name of the salt from the names of the acid and the metal or base. Working out which acid and base or metal you need to make a named salt can be a bit trickier. It is usually safest to use metal oxides.

We can make salts by reacting acids with metals or bases. Acids will react with metals that are above hydrogen in the reactivity series. These metals react with acids to form hydrogen gas and a salt.

$$ACID + METAL \rightarrow SALT + HYDROGEN$$

However, the reactions with the most reactive metals are too violent to be done safely.

Bases are metal oxides or metal hydroxides. They react with acids to form a salt and water.

$$ACID + BASE \rightarrow SALT + WATER$$

A metal, or a base that is insoluble in water, is added a little at a time to the acid until all of the acid has reacted. The mixture is then filtered to remove the excess solid, leaving a solution of the salt. The solid salt is made when water is evaporated from the solution so that it crystallises.

Chlorides are made from hydrochloric acid, nitrates from nitric acid and sulfates from sulfuric acid.

Key words: excess, chlorides, nitrates, sulfates

Making crystals of copper sulfate

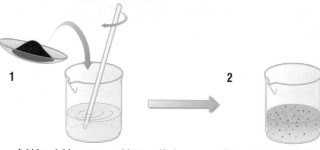

1

Add insoluble copper oxide to sulfuric acid and stir. Warm gently on a tripod and gauze (do not boil).

2

The solution turns blue as the reaction occurs, showing that copper sulfate is being formed

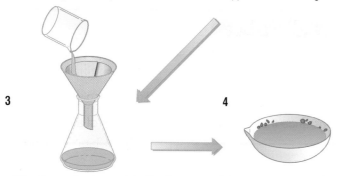

3

When the reaction is complete, filter the solution to remove excess copper oxide

4

We can evaporate the water so that crystals of copper sulfate are left

GET IT RIGHT!

Watch the spelling of hydrochloric acid, and remember that it forms chlorides.

CHECK YOURSELF

1 What are the products when an acid reacts with a base?

2 Why is an excess of the solid base added to the acid when making a salt?

3 Name the salt made from hydrochloric acid and zinc oxide.

4 Name the acid and base you would use to make copper nitrate.

Making salts from solutions

KEY POINTS

1 An indicator is used to find when an acid has exactly reacted with an alkali to form a salt.

2 We can make an insoluble salt by mixing two solutions containing the ions in the salt.

3 Ions can be removed from solutions by precipitation.

We can make soluble salts by reacting an acid and an alkali.

$$\text{ACID} + \text{ALKALI} \rightarrow \text{SALT} + \text{WATER}$$

We can summarise the reaction between any acid and alkali by just showing the ions that react:

$$H^+ (aq) \quad + \quad OH^- (aq) \quad \rightarrow \quad H_2O (l)$$
(from acid)   (from alkali)

However, there is no visible change when the solutions react so we need an indicator to show when the reaction is complete. The indicator can be removed from the solution after the reaction.

Alternatively, the volumes of acid and alkali needed to produce the salt are found using an indicator and these volumes of fresh solutions are mixed. The pure salt is obtained by crystallisation.

Ammonia solution is an alkali that does not contain a metal. It forms ammonium salts, such as ammonium nitrate, which are used as fertilisers.

We can make insoluble salts by mixing solutions of soluble salts that contain the ions needed. For example, we can make lead iodide by mixing solutions of lead nitrate and potassium iodide. The lead iodide forms a precipitate that can be filtered and dried.

Some pollutants can be removed from water as precipitates by adding ions that react with them to form insoluble salts.

Key words: ammonia, ammonium, precipitate

Making lead chloride

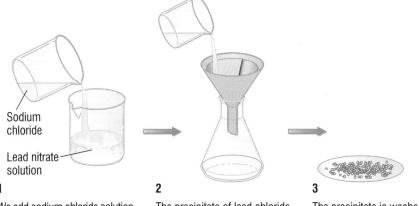

1
We add sodium chloride solution to lead nitrate solution and stir

2
The precipitate of lead chloride that forms is filtered off from the solution

3
The precipitate is washed with distilled water and dried

CHECK YOURSELF

1 Sodium sulfate can be made by reacting sodium hydroxide and sulfuric acid. What other substance will you need to add to the reaction mixture?

2 Name the salt formed when ammonia solution is reacted with hydrochloric acid.

3 Complete the equation for the reaction to make insoluble copper carbonate:

... sulfate + sodium ... → copper carbonate + sodium sulfate

C2 7   End of chapter questions

1 A solution has a pH value of 2. What does this tell you about the solution?

2 What is an alkali?

3 Name the products of the reaction between magnesium and sulfuric acid.

4 Name the products of the reaction between hydrochloric acid and copper oxide.

5 Why is an indicator needed when reacting an acid with an alkali?

6 Name the products when sodium hydroxide solution reacts with nitric acid.

7 Write an equation to show how the ions in an acid and alkali react so that it becomes neutral.

8 Suggest the name of a solution you could add to lead nitrate to produce insoluble lead sulfate.

1 Hydrogen peroxide decomposes to produce oxygen gas and water:

$$2H_2O_2(aq) \rightarrow 2H_2O(aq) + O_2(g)$$

The reaction is catalysed by manganese(IV) oxide.

Some students added 2g of manganese(IV) oxide to $20\,cm^3$ of hydrogen peroxide solution and measured the volume of gas produced. Their results are shown in the table.

Time (min)	0	1	2	3	4	5	6	7	8	9
Volume of gas (cm³)	0	18	34	48	59	65	74	78	80	80

(a) Plot a graph of the results with time on the horizontal axis and volume of gas on the vertical axis. Draw a smooth line through the points, omitting any result that is anomalous. (4 marks)

(b) (i) How can you tell from the graph that the rate of reaction was fastest at the start of this experiment? (1 mark)
    (ii) Explain, in terms of particles, why the rate of reaction was fastest at the start. (2 marks)

(c) The students repeated the experiment with 2g of manganese(IV) oxide that was more finely powdered. All other conditions were kept the same. Sketch a line on the same axes to show the results you would expect for this experiment. (2 marks)

2 The electrolysis of molten sodium chloride is used to produce sodium metal. The diagram shows the type of electrolysis cell that is used.

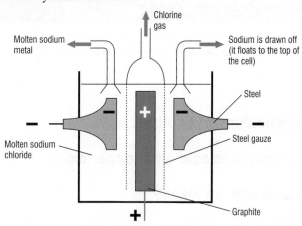

(a) Why must the sodium chloride be molten? (1 mark)

(b) Explain how sodium is produced at the negative electrode. (2 marks)

(c) Explain how chlorine is produced at the positive electrode. (2 marks)

(d) Suggest why the positive electrode is made of graphite. (1 mark)

(e) The electrolysis of aqueous sodium chloride solution (brine) is also done in industry. Hydrogen is produced at the negative electrode.
    (i) What are the other two products when brine is electrolysed? (2 marks)
    (ii) Explain why hydrogen is produced at the negative electrode. (2 marks)

3 The Haber Process is used to make ammonia.

(a) The table shows the percentage yield of ammonia at different temperatures and pressures.

Pressure (atmospheres)	Percentage yield of ammonia at 350°C (%)	Percentage yield of ammonia at 500°C (%)
50	25	5
100	37	9
200	52	15
300	63	20
400	70	23
500	74	25

    (i) Draw graphs of this data on the same axes. Put the percentage yield of ammonia (%) on the vertical axis and pressure (atmospheres) on the horizontal axis. Plot the points and draw a smooth line for each temperature. Label each line with its temperature. (4 marks)
    (ii) Use your graphs to find the conditions needed to give a yield of 30% ammonia. (1 mark)
    (iii) On the same axes, sketch the graph you would expect for a temperature of 450°C. (1 mark)

(b) This equation represents the reaction in which ammonia is formed:

$$N_2(g) + 3H_2(g) \rightleftharpoons 2NH_3(g)$$

    (i) What does the symbol $\rightleftharpoons$ in this equation tell you about the reaction? (1 mark)
    (ii) Explain why a temperature of 450°C is used in industrial processes to make ammonia. (2 marks)
    (iii) Explain why a pressure of 200 atmospheres is used in industrial processes to make ammonia. (2 marks)

[Higher]

 Test & Assessment Interactive quizzes, answers and hints online!

The answer is worth 4 out of the 5 marks available.

The responses worth a mark are underlined in red.

We can improve the answer in several ways:

Explain as fully as you can why hydrogen is produced at the negative electrode when a solution of sodium chloride in water is electrolysed. *(5 marks)*

Sodium chloride solution contains Na⁺ ions and H⁺ ions. The sodium ions come from sodium chloride and the hydrogen ions from water. Both these positive ions are attracted to the negative electrode but only the hydrogen ions are reduced because they are less reactive than sodium ions. The reaction that takes place at the electrode is H⁺ + e⁻ → H.

The answer is clearly expressed and begins well. However, the reason that hydrogen ions are reduced in preference to sodium ions is incorrect. It is sodium metal that is more reactive than hydrogen, and therefore sodium ions are more difficult to reduce.

It is a good idea to include equations for reactions when possible, but the reaction here is incomplete and shows only the formation of hydrogen atoms, not hydrogen gas, which is made of molecules of H_2. However, if reduction had not been mentioned it could have gained the mark for reducing hydrogen ions.

The answer is worth 4 marks out of the 7 marks available.

The responses worth a mark are underlined in red.

We can improve the answer in several ways:

Antacid tablets contain calcium carbonate. They neutralise excess acid in the stomach.

(a) How does the pH in the stomach change when the tablets 'neutralise excess acid'? *(1 mark)*

(b) Write an equation to show the reaction between the ions in the neutralisation reaction. *(2 marks)*

(c) Chewing the tablet cures indigestion faster than swallowing the tablet in one piece. Explain why, as fully as you can, using particle theory. *(4 marks)*

(a) The pH gets less.

(b) $H^+ + OH^- \rightarrow H_2O$

(c) Chewing the tablet breaks it down into smaller pieces. The smaller pieces have more surface area, so the acid can get to them more easily. Particles of acid bump into the bits of tablet and when they do this they react. Because the bits of tablet are smaller the indigestion is cured faster.

This is incorrect and scores no mark. The student has confused acidity with pH. The acidity decreases ('gets less'), but the pH increases when the acid is neutralised.

The answer starts well, although 'get to them more easily' is not the same as 'a faster reaction'.

The increase in frequency of collisions and the increase in rate of reaction are not made clear.

This sentence just repeats information from earlier in the answer. Replace with 'Particles of acid bump into the bits of tablet more frequently and so the rate of the reaction increases.'

Further chemistry (Chapters 1–3)

Checklist

This spider diagram shows the topics in the unit. You can copy it out and add your notes and questions around it, or cross off each section when you feel confident you know it for your exams.

Tick when you:			
reviewed it after your lesson	☑	☐	☐
revised once – some questions right	☑	☑	☐
revised twice – all questions right	☑	☑	☑
Move on to another topic when you have all three ticks.			

Chapter 1 Development of the periodic table

1.1	The early periodic table	☐	☐	☐
1.2	The modern periodic table	☐	☐	☐
1.3	Group 1 – the alkali metals	☐	☐	☐
1.4	Group 7 – the halogens	☐	☐	☐
1.5	The transition elements	☐	☐	☐

Chapter 2 More about acids and bases

2.1	Strong and weak acids/alkalis	☐	☐	☐
2.2	Titrations	☐	☐	☐
2.3	Titration calculations	☐	☐	☐

Chapter 3 Water

3.1	Water and solubility	☐	☐	☐
3.2	Solubility curves	☐	☐	☐
3.3	Hard water	☐	☐	☐
3.4	Removing hardness	☐	☐	☐
3.5	Water treatment	☐	☐	☐

What are you expected to know?

Chapter 1 Development of the periodic table (See students' book pages 210–221)

- In the 19th century scientists tried to classify elements. They arranged them in order of atomic weight and produced the first periodic tables.
- The early tables were incomplete and some elements were put into groups where their properties did not match the other elements.
- Mendeleev left gaps for undiscovered elements and produced a better table.
- The modern periodic table is based on atomic (proton) numbers and the arrangement of electrons in atoms of the elements.
- Elements in the same group have the same number of electrons in their highest occupied energy level (outer shell).
- The higher the energy level the more easily electrons are lost and the less easily they are gained. This explains why reactivity increases down Group 1 but decreases down Group 7. [Higher Tier only]
- The elements in Group 1 are all metals that react with water, form alkalis and form ions with a single + charge.
- The elements in Group 7 are all non-metals, have molecules with two atoms, have coloured vapours, and form ions with a single – charge.
- Transition elements are all metals that are stronger, harder, less reactive and (except for mercury) have higher melting points than the metals in Group 1.
- Transition elements form ions with different charges, form coloured compounds and are useful as catalysts because of their electronic structures.

19 F 9
35 Cl 17
80 Br 35
127 I 53
210 At 85

The Group 7 elements

Chapter 2 More about acids and bases (See students' book pages 224–231)

- Acids produce hydrogen ions in aqueous solutions.
- Acids are proton donors and bases are proton acceptors. [Higher Tier only]
- Alkalis are soluble bases that produce hydroxide ions in aqueous solutions.
- Strong acids and alkalis are completely ionised in aqueous solutions.
- Weak acids and alkalis only partly ionise in aqueous solutions.
- Titrations can be used to find reacting volumes and concentrations of solutions of acids and alkalis if a suitable indicator is used.

1 mol/dm³   1 mol/dm³

Weak acid   Strong acid

Chapter 3 Water (See students' book pages 234–245)

- The water cycle involves evaporation and condensation.
- Many ionic compounds and some molecular substances dissolve in water.
- The solubility of most solids increases as the temperature increases.
- A solution is saturated when it cannot dissolve any more solute.
- The solubility of many gases decreases as the temperature increases.
- Hard water contains compounds, usually of calcium or magnesium, that have dissolved and that react with soap to form scum.
- Hard water can increase costs because it uses more soap and it can form scale when it is heated.
- Hard water can be made soft by removing the dissolved calcium or magnesium ions using sodium carbonate or ion exchange.
- Good quality drinking water is essential for humans.
- Water quality can be improved by filtration, sterilisation, and ion exchange.
- Pure water can be produced by distillation.

1. How did scientists in the 19th century try to classify the elements?

2. Who produced a periodic table with gaps for elements that had not then been discovered?

3. In the modern periodic table, how are the electronic structures of elements in a group the same?

4. How are the electronic structures of elements in a group different?

5. How does the reactivity of the elements change going down Group 1?

6. Why does the reactivity in Group 1 change as described in question 5? [Higher Tier only]

7. Make a list of facts about chlorine that show it is a halogen.

8. What is the trend in reactivity in the halogen group?

9. Why are transition elements useful as building materials?

10. Why are there ten transition elements in each period? [Higher Tier only]

---

students' book page 210

C3 1.1 The early periodic table

KEY POINTS

1. When scientists arranged elements in order of their atomic weights they found similar properties at regular intervals.
2. Mendeleev produced a table that became the basis for the modern periodic table.

During the 19th century, many scientists tried to find ways to classify elements based on their properties and atomic weights.

In 1863 Newlands proposed his *law of octaves*, which stated that similar properties were repeated every eighth element:

- He put the 56 known elements into seven groups according to their atomic weights.

GET IT RIGHT!

The early tables of elements were based on atomic weights (relative atomic masses).

EXAM HINTS

Remember that the scientists in the 19th century knew nothing about protons and electrons.

EXAMINER SAYS...

You do not need to know the details of the early tables of elements. In the examination, questions will contain any information and tables of elements you need to answer the questions.

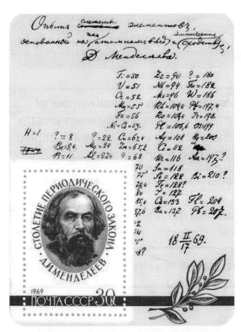

Some of Mendeleev's original notes on the periodic table, together with a Russian stamp issued in his honour in 1969

- However, their properties did not match very well within the groups and so other scientists did not accept his ideas.

In 1869 Mendeleev produced a better table by leaving gaps for undiscovered elements:

- When some of the missing elements were discovered, they were found to have properties as predicted by Mendeleev. Then other scientists more readily accepted his ideas.
- Mendeleev's table became the basis for the modern periodic table.

Key words: atomic weights, octaves, periodic table

CHECK YOURSELF

1 What was Newlands' *law of octaves*?

2 Why did other scientists not accept Newlands' law?

3 What was Mendeleev's inspirational idea that improved his table?

4 What evidence made Mendeleev's ideas more acceptable?

---

students' book page 212 **C3 1.2** **The modern periodic table**

KEY POINTS

1 The modern periodic table is based on atomic numbers.

2 Elements within a group have similar chemical properties.

3 The reactivity of metals increases going down a group.

4 The reactivity of non-metals decreases going down a group.

Scientists found out about protons and electrons at the start of the 20th century. Soon after this, they developed models of the arrangement of electrons in atoms. The elements were arranged in the periodic table in order of their atomic numbers. They were lined up in groups.

The groups of elements have similar chemical properties because their atoms have the same number of electrons in their highest occupied energy level (outer shell).

GET IT RIGHT!

- The most reactive metals are at the *bottom* of their group.
- The most reactive non-metals are at the *top* of their group.

Group numbers

1	2										3	4	5	6	7	0
					Atomic mass → 1 **H** Atomic number → 1											4 **He** 2
7 **Li** 3	9 **Be** 4										11 **B** 5	12 **C** 6	14 **N** 7	16 **O** 8	19 **F** 9	20 **Ne** 10
23 **Na** 11	24 **Mg** 12										27 **Al** 13	28 **Si** 14	31 **P** 15	32 **S** 16	35.5 **Cl** 17	40 **Ar** 18
39 **K** 19	40 **Ca** 20	45 **Sc** 21	48 **Ti** 22	51 **V** 23	52 **Cr** 24	55 **Mn** 25	56 **Fe** 26	59 **Co** 27	59 **Ni** 28	63.5 **Cu** 29	65 **Zn** 30	70 **Ga** 31	73 **Ge** 32	75 **As** 33	79 **Se** 34	80 **Br** 35	84 **Kr** 36
85 **Rb** 37	88 **Sr** 38	89 **Y** 39	91 **Zr** 40	93 **Nb** 41	96 **Mo** 42	98 **Tc** 43	101 **Ru** 44	103 **Rh** 45	106 **Pd** 46	108 **Ag** 47	112 **Cd** 48	115 **In** 49	119 **Sn** 50	122 **Sb** 51	128 **Te** 52	127 **I** 53	131 **Xe** 54
133 **Cs** 55	137 **Ba** 56	139 **La** 57	178 **Hf** 72	181 **Ta** 73	184 **W** 74	186 **Re** 75	190 **Os** 76	192 **Ir** 77	195 **Pt** 78	197 **Au** 79	201 **Hg** 80	204 **Tl** 81	207 **Pb** 82	209 **Bi** 83	210 **Po** 84	210 **At** 85	222 **Rn** 86
223 **Fr** 87	226 **Ra** 88	227 **Ac** 89														

Elements 58–71 and 90–103 (all metals) have been omitted

The modern periodic table

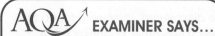

AQA EXAMINER SAYS...

You may describe electron arrangements in terms of energy levels or shells. It is generally accepted that 'outer electrons' of an atom refer to the electrons in the outer shell or highest occupied energy level.

Within a group the reactivity of the elements depends on the *total number* of electrons. Going down a group, there are more occupied energy levels and the atoms get larger. As the atoms get larger, the electrons in the highest occupied energy level (outer shell) are less strongly attracted by the nucleus.

- When metals react they *lose* electrons, so the reactivity of metals in a group *increases* going *down* the group.
- When non-metals react they *gain* electrons, so the reactivity of non-metals *decreases* going *down* a group.

Key words: atomic number, highest occupied energy level, outer shell

EXAM HINTS

Diagrams can help with explanations and may save lots of words.

CHECK YOURSELF

1 Why do elements in a group in the periodic table have similar chemical properties?

2 Which is more reactive, lithium or caesium?

3 Which is more reactive, chlorine or bromine?

4 Why do metals get more reactive going down a group? [Higher Tier only]

KEY POINTS

1 The elements in Group 1 are all metals that form positive ions with a charge of 1+. They react with water to produce hydrogen and an alkali.

2 Going down the group: reactivity increases, melting and boiling points decrease.

GET IT RIGHT!

The alkali metals form only ionic compounds in which their ions have a single positive charge.

BUMP UP YOUR GRADE

Grade A candidates should be able to explain the properties of the alkali metals in terms of their electron arrangements.

- The Group 1 elements are reactive metals.
- They are soft solids at room temperature with low melting and boiling points that decrease going down the group.
- They have low densities, so lithium, sodium and potassium float on water.
- They all react readily with air and water. With water they produce hydrogen gas and a metal hydroxide that is a strong alkali, e.g.

$$\text{lithium} + \text{water} \rightarrow \text{lithium hydroxide} + \text{hydrogen}$$
$$2Li(s) + 2H_2O(l) \rightarrow 2LiOH(aq) + H_2(g)$$

- They all have one electron in their highest occupied energy level (outer shell). They lose this electron in reactions to form ionic compounds in which their ions have a single positive charge, e.g. Na^+.
- The Group 1 elements get more reactive going down the group.

- Reactivity increases going down Group 1 because the outer electron is less strongly attracted to the nucleus as the number of occupied energy levels increases and the atoms get larger.

HIGHER

Lithium and potassium reacting with water (the lithium is on the right of the trough)

- They react with the halogens to form salts that are white or colourless crystals, e.g.

$$\text{sodium} + \text{chlorine} \rightarrow \text{sodium chloride}$$
$$2Na(s) + Cl_2(g) \rightarrow 2NaCl(s)$$

These compounds dissolve in water to form colourless solutions.

Key words: alkali metals, ionic compounds, halogens

AQA EXAMINER SAYS…

You should be able to describe the properties of lithium, sodium and potassium and their reactions with water and chlorine. Questions may contain other information that you are expected to use, applying your knowledge to make predictions about the members of the group that you have not seen in the laboratory.

CHECK YOURSELF

1 Why are the Group 1 elements called 'alkali metals'?

2 What is unusual about lithium, sodium and potassium compared with other metals when they are put into water?

3 Name and give the formula of the compound formed when lithium reacts with bromine.

4 Why is potassium more reactive than sodium? [Higher Tier only]

---

students' book page 216 | **C3 1.4** **Group 7 – the halogens**

KEY POINTS

1 The elements in Group 7 are all non-metals with low melting and boiling points.

2 They form compounds with metals and non-metals.

3 The reactivity of the halogens decreases going down the group.

AQA EXAMINER SAYS…

Make sure you revise ionic and covalent bonding so that you are clear about the differences between ionic compounds and molecular substances.

GET IT RIGHT!

Halogens form ionic compounds with metals and covalently bonded compounds with non-metals.

- The halogens are non-metallic elements in Group 7 of the periodic table.
- They exist as small molecules made up of pairs of atoms.
- They have low melting and boiling points that increase down the group.
- At room temperature fluorine is a pale yellow gas, chlorine is a green gas, bromine is a red-brown liquid and iodine is a grey solid. Iodine easily vaporises to a violet gas.
- All of the halogens have seven electrons in their highest occupied energy level.
- They form ionic compounds with metals in which their ions have a charge of –1.
- They also bond covalently with other non-metals, forming molecules.
- The reactivity of the halogens decreases going down the group.

- Their reactivity decreases going down the group because attraction of the outer electrons to the nucleus decreases as the number of occupied energy levels increases. **HIGHER**

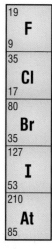

The Group 7 elements

75

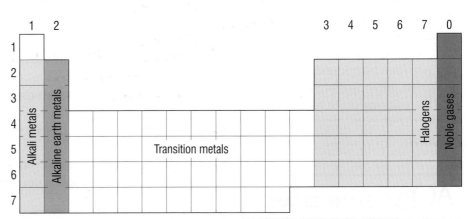

Grade A candidates should be able to write balanced equations for all of the reactions in this chapter.

EXAM HINTS

Fluorine reacts **f**uriously – it is the most reactive halogen.

- A more reactive halogen is able to displace a less reactive halogen from its compounds in solution.

Key words: halogens, non-metallic, molecules, ionic, covalent

CHECK YOURSELF

1 Why do the halogens have low melting and boiling points?

2 Why do halogens form both ionic and covalent compounds?

3 Why is fluorine the most reactive halogen? [Higher Tier only]

4 How could you show that chlorine is more reactive than bromine?

students' book page 218

C3 1.5 The transition elements

KEY POINTS

1 The elements between Groups 2 and 3 in the periodic table are called the 'transition elements'.

2 Transition elements are all metals with high densities.

3 Many transition metals form several different ions and compounds that are coloured.

4 Transition metals and their compounds are useful as catalysts.

GET IT RIGHT!

All transition elements are metals that form positive ions with various charges. For example, copper can form Cu^+ and Cu^{2+} ions.

- The transition elements are found in the periodic table between Groups 2 and 3.
- They are all metals and, except for mercury, have higher melting and boiling points than the alkali metals.
- Most are strong and dense and are useful as building materials, often as alloys.
- They are malleable and ductile.
- They are good conductors of heat and electricity.
- They react only slowly, or not at all, with oxygen and water at ordinary temperatures.

The transition elements lie between Groups 2 and 3

AQA EXAMINER SAYS...

Grade A candidates should be able to link their knowledge and understanding of this section with related topics in other units of the specification.

EXAM HINTS

Use the periodic table on the Data Sheet to identify unfamiliar transition elements that may be named in questions. (See page 114.)

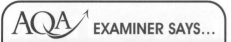
- They form positive ions with various charges, e.g. Fe^{2+} and Fe^{3+}.

- The different charges on their ions arise mainly because they have a small number of electrons in their highest occupied energy level and an incomplete lower energy level. This lower energy level can hold up to ten more electrons for elements beyond calcium, and so there are ten transition elements in each period.

- Compounds of transition metals are often brightly coloured.
- Many transition metals and their compounds are catalysts for chemical reactions.

Key words: transition elements, catalyst, alloy

CHECK YOURSELF

1 Where are the transition metals in the periodic table?

2 Give two ways in which transition metals are similar to the elements in Group 1.

3 Give three ways in which transition metals are different to the elements in Group 1.

4 What feature of their electronic structures gives transition metals their special properties? [Higher Tier only]

C3 1   End of chapter questions

1 Why did Newlands put the elements into seven groups?

2 Explain why Mendeleev's table was better than Newlands' 'octaves'.

3 What discoveries at the start of the 20th century led to the modern periodic table?

4 Why does the reactivity of elements in a group change going down the group? [Higher Tier only]

5 a) Write a word equation for the reaction of sodium metal with water.
   b) Now write a symbol equation for the same reaction. [Higher Tier only]

6 Why is lithium the least reactive alkali metal?

7 Why do the halogens have low melting and boiling points?

8 a) Write a word equation for the reaction of chlorine with potassium bromide.
   b) Now write a symbol equation for the same reaction. [Higher Tier only]

9 Copper forms two chlorides, CuCl and $CuCl_2$. What does this tell you about copper?

10 Predict two physical and two chemical properties of the element molybdenum, Mo.

1. What are strong acids?

2. How is a solution of ethanoic acid different from a solution of nitric acid with the same concentration?

3. What is a titration?

4. Why is an indicator needed in a titration of an acid and an alkali?

5. What can be calculated from the results of a titration? [Higher Tier only]

6. How can you calculate the concentration of a substance in solution in grams per cubic decimetre (dm³) from a concentration in moles per dm³? [Higher Tier only]

---

students' book
page 224

C3 2.1   Strong and weak acids/alkalis

KEY POINTS

1 Strong acids and alkalis ionise completely in aqueous solutions.
2 Weak acids and alkalis only partly ionise in aqueous solutions.
3 Acids are proton donors and bases are proton acceptors. [Higher Tier only]

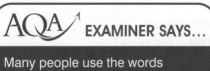

EXAMINER SAYS...

Many people use the words 'strong' and 'weak' to describe concentrations of solutions, but in chemistry these words are used to describe different types of acid and alkali. It is best to use 'strong' and 'weak' only when describing acids and alkalis that ionise completely or partially.

When acids dissolve in water they produce hydrogen ions in the solution:

● Hydrogen ions are protons.
● In aqueous solutions, water molecules surround the protons to keep them in solution. This is called 'hydration'.
● We represent hydrated protons as $H^+(aq)$.
● Acids are proton donors. [Higher Tier only]
● Strong acids, such as hydrochloric, sulfuric and nitric acids, ionise completely when they dissolve in water.
● Acids that only partly ionise in water, such as ethanoic acid and citric acid, are called 'weak acids'.
● Solutions of weak acids have higher pH values and react more slowly than solutions of strong acids with the same concentration.

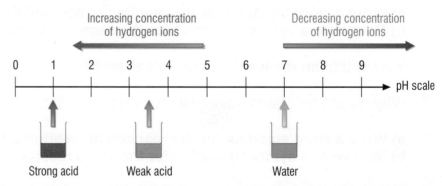

When we compare two acids with the same concentration, the *stronger* acid will have the *lower* pH

GET IT RIGHT!

Solutions of acids and alkalis can be dilute or concentrated, but it is the particular acid or alkali that is strong or weak.

Alkalis are soluble bases that produce hydroxide ions $OH^-(aq)$ in solution.

- In aqueous solutions strong alkalis, such as sodium hydroxide, are completely ionised.
- Weak alkalis, such as ammonia, only partly ionise in aqueous solution.
- Solutions of weak alkalis have lower pH values than solutions of strong alkalis with the same concentration.
- Bases react with protons and, so, are called 'proton acceptors'. [Higher Tier only]

Key words: hydrated protons, proton donor, proton acceptor, ionise

CHECK YOURSELF

1 What does $H^+(aq)$ represent?

2 What type of acid is dilute hydrochloric acid?

3 Name two strong alkalis.

4 How can you use pH to tell a weak alkali from a strong one?

---

students' book page 226 ## C3 2.2 Titrations

KEY POINTS

1 Titrations can be used to find the volumes of acids and alkalis that react completely.

2 A suitable indicator is used to show the end point of a titration.

GET IT RIGHT!

The end point of the titration is when the acid and alkali have reacted completely.

BUMP UP YOUR GRADE

Make sure you can describe how to use a pipette and a burette to obtain results that are precise and reliable in titrations.

AQA EXAMINER SAYS...

The solution is neutral at the end point and the pH is 7 only if the acid and alkali are both strong. For a strong acid and weak alkali the end point is below pH 7 and for a strong alkali and weak acid the end point is above pH 7.

When solutions of acids and alkalis react to form salts and water a neutralisation reaction takes place. The volumes of solutions that react exactly can be found by using a titration. A pipette is used to accurately measure a volume of alkali that is put into a conical flask. An indicator is added and the acid solution is added gradually from a burette.

When the indicator changes colour the end point has been reached. The volume of acid used is found from the burette readings. The titration should be repeated several times to improve the reliability of the result.

To obtain a sharp, distinct end point one of the solutions used in a titration must be a strong acid or alkali. The indicator that is used depends on the strengths of the acid and alkali:

- strong acid + strong alkali – use any acid–base indicator

HIGHER
- weak acid + strong alkali – use phenolphthalein
- strong acid + weak alkali – use methyl orange.

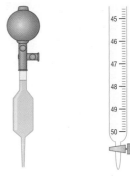

A pipette and a burette

Key words: neutralisation, titration, indicator, end point

CHECK YOURSELF

1 What is used to measure the known volume of solution that is put into the conical flask in a titration?

2 Why should you repeat the titration several times?

3 Why should one of the solutions be a strong acid or alkali?

4 Which indicator would you use for ethanoic acid and sodium hydroxide? [Higher Tier only]

EXAMINER SAYS...

Titrations are an important method of quantitative analysis. The information can be used to find concentrations, amounts, purities and formulae of substances. Practise as many calculations as possible.

Titrations are used to find the volumes of solutions that react exactly:

- If the concentration of one of the solutions is known, and the volumes that react together are known, the concentration of the other solution can be calculated.
- This information can be used to find the amount of a substance in a sample. The concentrations are calculated using balanced equations and moles.

Worked example

A student put 25.0 cm³ of sodium hydroxide solution with an unknown concentration into a conical flask using a pipette. The sodium hydroxide reacted with exactly 20.0 cm³ of 0.50 mol/dm³ hydrochloric acid added from a burette. What was the concentration of the sodium hydroxide solution?

Solution

The equation for this reaction is:

$$NaOH \ (aq) + HCl \ (aq) \rightarrow NaCl \ (aq) + H_2O \ (l)$$

This equation tells us that 1 mol of NaOH reacts with 1 mol of HCl.

The concentration of the HCl is 0.50 mol/dm³, so

0.50 moles of HCl are dissolved in 1000 cm³ of acid, and

$\dfrac{0.50}{1000}$ moles of HCl are dissolved in 1 cm³ of acid, therefore

$\dfrac{0.50}{1000} \times 20.0$ moles of HCl are dissolved in 20.0 cm³ of acid.

There are 0.010 moles of HCl dissolved in 20.0 cm³ of acid.

The equation for the reaction tells us that 0.010 moles of HCl will react with exactly 0.010 moles of NaOH. This means that there must have been 0.010 moles of NaOH in the 25.0 cm³ of solution in the conical flask. To calculate the concentration of NaOH in the solution in the flask we need to calculate the number of moles of NaOH in 1 dm³ (1000 cm³) of solution.

0.010 moles of NaOH are dissolved in 25.0 cm³ of solution, so

$\dfrac{0.010}{25}$ moles of NaOH are dissolved in 1 cm³ of solution, and there are

$\dfrac{0.010}{25} \times 1000 = 0.40$ moles of NaOH in 1 dm³ (1000 cm³) of solution.

The concentration of the sodium hydroxide solution is 0.40 mol/dm³.

- The mole quantities can be converted into masses. That's because a mole is the relative formula mass of a substance in grams.

EXAM HINTS

number of moles =

$$\frac{\text{mass in grams}}{\text{relative formula mass}}$$

 EXAMINER SAYS…

Top-grade students should be able to balance equations for reactions, calculate amounts of substances from titration data and apply these skills to solve problems in different contexts.

GET IT RIGHT!

Make sure you check the number of moles that react from the balanced equation.

- Using the relative formula mass of NaOH, you can convert the concentration of the sodium hydroxide in moles per dm^3 into grams per dm^3 (e.g. $40 \times 0.40 = 16 \, g/dm^3$).
- You can then calculate the mass of sodium hydroxide needed to make any volume of solution with the same concentration (e.g. to make $100 \, cm^3$ of solution you will need $1.6 \, g$ of sodium hydroxide).

Key words: mol per dm^3, g per dm^3, mole, relative atomic mass

CHECK YOURSELF

1 What information do you need to calculate the concentration of a solution of nitric acid that was titrated with sodium hydroxide solution?

2 $25.0 \, cm^3$ of a solution of potassium hydroxide reacted exactly with $12.5 \, cm^3$ of sulfuric acid with concentration of $0.1 \, mol$ per dm^3. What was the concentration of the potassium hydroxide solution in:
(a) mol per dm^3,
(b) g per dm^3?

C3 2   End of chapter questions

1 **Why is dilute nitric acid a strong acid?**

2 **Describe one way that you could show that ethanoic acid is a weak acid.**

3 **Name the two pieces of apparatus used to measure volumes in titrations and state which one has a tap.**

4 **Why should citric acid solution and ammonia solution not be used in a titration?**

5 **$25.0 \, cm^3$ of hydrochloric acid solution reacted with $12.5 \, cm^3$ of sodium hydroxide solution. What can you deduce from this information? [Higher Tier only]**

6 **What volume of nitric acid with concentration $0.10 \, mol$ per dm^3 would exactly neutralise $20 \, cm^3$ of sodium hydroxide solution with a concentration of $4.0 \, g$ per dm^3? [Higher Tier only]**

① **Explain how the water cycle produces fresh water.**

② **What is a saturated solution?**

③ **What are solubility curves?**

④ **How does the solubility of most gases change with temperature?**

⑤ **What makes water hard?**

⑥ **What are the disadvantages of hard water?**

⑦ **How does washing soda soften hard water?**

⑧ **What is ion exchange?**

⑨ **What do we mean by 'water fit to drink'?**

⑩ **How is water made fit to drink?**

---

students' book
page 234
C3 3.1   Water and solubility

KEY POINTS

1 Many ionic compounds and some molecular substances dissolve in water.
2 A saturated solution has dissolved the maximum amount of solute at a given temperature.

EXAMINER SAYS…

Make sure you know the processes in the water cycle and that you can explain how it produces fresh water.

Water is constantly cycled around the Earth and its atmosphere:

● When water evaporates it produces water vapour that mixes with air.
● When the water vapour cools it condenses into droplets forming mist and clouds. Large droplets fall as rain and this provides fresh water.
● Many gases, including oxygen and carbon dioxide, are soluble in water and dissolve in rain.
● Dissolved oxygen is essential for aquatic life.

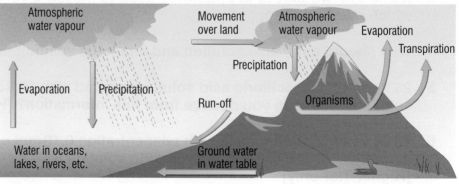

The water cycle

Water vapour cannot be seen. It is only when it has condensed to form droplets that it appears as mist or clouds.

Many ionic compounds are soluble, but many covalent compounds do not dissolve in water:

- Water that has been in contact with the ground may contain dissolved solids.
- Solubility is usually measured in grams of solute per 100 grams of solvent at a given temperature.
- A saturated solution contains the maximum amount of solute that will dissolve at the temperature of the solution. Cooling a hot saturated solution of a solid causes some of the solute to crystallise.

Key words: water vapour, solubility, solute, solvent, saturated solution

EXAM HINTS

Use scientific terms in your answers to questions. Make sure you use 'solute', 'solvent', 'solution', 'soluble', 'saturated', 'vapour', 'mist', 'cloud', 'evaporate' and 'condense' correctly.

CHECK YOURSELF

1 What are the two main processes in the water cycle?

2 What types of substance dissolve in water?

3 What units are used when measuring solubility?

4 What usually happens when a hot saturated solution of a salt is cooled?

---

students' book
page 236

C3 3.2 Solubility curves

KEY POINTS

1 Solubility curves show how solubility changes with temperature.

2 The solubility of most solids increases with increasing temperature.

3 The solubility of gases decreases with increasing temperature.

AQA EXAMINER SAYS...

In the examinations you will be given solubility data to interpret. Then you will be asked to apply the information to solving problems.

EXAM HINTS

Temperature should be always on the x-axis for solubility curves.

The solubility of a substance at different temperatures can be plotted as a line graph, known as a solubility curve:

- Solubility curves can be used to find how much solute will dissolve at a particular temperature.
- They can also be used to find how much solute will separate out of a saturated solution for a given change in temperature.
- The solubility of most solids increases as temperature increases.

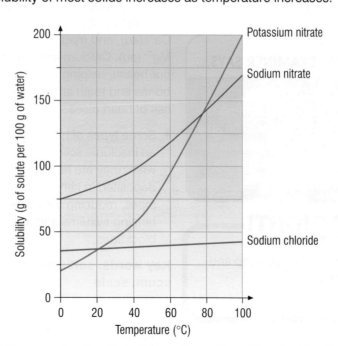

These solubility curves show the effect of temperature on three different solutes dissolving in water

GET IT RIGHT!

Most solids are more soluble at higher temperatures, but most gases are less soluble as the temperature increases.

BUMP UP YOUR GRADE

For top grades, you should be able to calculate the amount of solute that crystallises out of solution for given temperature changes.

The solubility of gases in water:

- Decreases as temperature increases. At higher temperatures less oxygen can dissolve in water. This affects fish and other aquatic life.
- Increases as pressure increases. Carbonated water and fizzy drinks contain carbon dioxide dissolved under high pressure. The dissolved gas comes out of solution as bubbles at ordinary pressure.

Key words: solubility curve, carbonated water

CHECK YOURSELF

1 How does the solubility of most solids change with temperature?

2 In what three ways can solubility curves be used?

3 Why are high temperatures a problem for some fish in ponds?

4 What is carbonated water?

KEY POINTS

1 Dissolved salts, usually of magnesium and calcium, cause hard water.
2 Soap reacts with the dissolved salts to form scum.
3 When hard water is heated it may form scale.

AQA EXAMINER SAYS...

You should know the main advantages and disadvantages of hard water and you should be able to evaluate information you are given about different samples of water.

GET IT RIGHT!

'Scum' is produced by soap and 'scale' by heat.

- Water that lathers easily with soap is said to be 'soft'.
- Hard water uses more soap to produce a lather and to wash effectively.
- Dissolved salts in hard water react with soap to form insoluble solids called 'scum'.
- The salts that react with soap are usually compounds of calcium or magnesium.

When water is in contact with rocks some compounds dissolve to produce a solution that contains calcium ions, Ca^{2+}(aq), and magnesium ions, Mg^{2+}(aq). Calcium ions are good for our health, helping to develop strong bones and teeth and also reducing the risk of heart disease.

- Some types of hard water produce an insoluble solid called 'scale' when they are heated. Scale can be deposited in kettles, boilers and pipes, reducing the efficiency of heating systems and causing blockages.

Key words: soft water, hard water, scum, scale

As scale builds up in heating systems and kettles, it not only makes them less efficient – it can stop them working completely

CHECK YOURSELF

1 What makes water hard?

2 How does hardness get into water?

3 Why is hard water good for you?

4 What is scale?

C3 3.4 Removing hardness

KEY POINTS

1 Removing dissolved calcium and magnesium ions softens hard water.
2 Two main methods are adding washing soda or using ion exchange.

AQA EXAMINER SAYS...

You may find other methods that can be used to soften water. They usually use principles similar to these two methods. Distilled water is soft, but distillation is a very expensive way to soften water and so it is not usually used.

GET IT RIGHT!

Precipitation removes ions from solution so they cannot react with soap.

Hard water can be made soft by removing the ions that react with soap. This can be done by precipitating out the ions or by replacing them with ions that do not react with soap.

A simple method is to add washing soda, which is sodium carbonate. The carbonate ions react with the ions that cause hardness forming precipitates. As they are no longer in solution the ions cannot react with soap. The carbonates of metals, except for those in Group 1, are insoluble.

Ion-exchange resins for water softening usually contain sodium ions. When hard water is passed through the resin, calcium and magnesium ions become attached to the resin and sodium ions take their place in the water. Sodium ions do not react with soap and so the water is soft. The sodium ions can be replaced in the resin by washing it with a concentrated solution of sodium chloride.

Key words: washing soda, ion-exchange resin

Washing soda is a simple way to soften water without the need for any complicated equipment

EXAM HINTS

Metal ions with a charge of 2+ or 3+ react with soap and also form insoluble carbonates.

GET IT RIGHT!

Describe hardness of water in terms of ions rather than compounds.

CHECK YOURSELF

1 What is soft water?

2 What are the two main methods of softening water?

3 Which metal carbonates are insoluble?

4 Explain how a sodium ion-exchange resin can be recharged.

C3 3.5 Water treatment

AQA EXAMINER SAYS...

You should know the main steps in water treatment and you should have considered some of the social and environmental aspects of water quality and supply.

Water is important for drinking, for washing and as a raw material. Water supplies need to be fit for these purposes. Drinking water should contain only very low, and therefore insignificant, levels of harmful substances.

Water from an appropriate source can be treated to make it safe to drink. This can be done by sedimentation and filtration to remove solids and by killing any microorganisms in the water. Chlorine is often used to kill bacteria in drinking water. Boiling the water will also destroy most microorganisms.

Good, clean water is a precious resource. Those of us lucky enough to have it can too easily take it for granted.

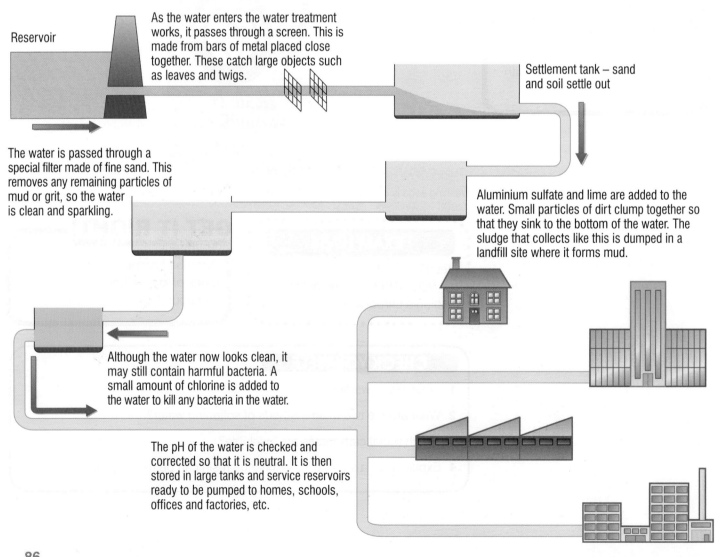

Reservoir

As the water enters the water treatment works, it passes through a screen. This is made from bars of metal placed close together. These catch large objects such as leaves and twigs.

Settlement tank – sand and soil settle out

The water is passed through a special filter made of fine sand. This removes any remaining particles of mud or grit, so the water is clean and sparkling.

Aluminium sulfate and lime are added to the water. Small particles of dirt clump together so that they sink to the bottom of the water. The sludge that collects like this is dumped in a landfill site where it forms mud.

Although the water now looks clean, it may still contain harmful bacteria. A small amount of chlorine is added to the water to kill any bacteria in the water.

The pH of the water is checked and corrected so that it is neutral. It is then stored in large tanks and service reservoirs ready to be pumped to homes, schools, offices and factories, etc.

Water filters can be used to improve the taste of water. They often contain charcoal and an ion-exchange resin with silver or another substance to prevent the growth of bacteria.

Pure water can be made either by distillation or by de-ionisation. De-ionising uses an ion-exchange column to remove all ions except H^+ and OH^- from water.

Key words: sedimentation, pure, de-ionisation

CHECK YOURSELF

1 What type of substance may be present in drinking water only at an insignificant level?

2 Why is chlorine added to drinking water?

3 What is usually in the filter cartridges used to improve the taste of water?

4 What is de-ionisation?

C3 3   End of chapter questions

1 **Name two gases that may be dissolved in rainwater.**

2 **Complete the sentence: A saturated solution contains . . . .**

3 **Explain how you would use a solubility curve to find the mass of crystals that would separate from a solution that was cooled from 60°C to 20°C.**

4 **How is carbonated water made?**

5 **Why does hard water produce scum?**

6 **What are the problems caused by scale?**

7 **Why does the precipitation of metal carbonates make hard water soft?**

8 **Explain how an ion-exchange resin softens hard water.**

9 **What are the main steps in providing drinking water?**

10 **What methods can be used to make pure water?**

1 Newlands and Mendeleev used relative atomic masses to put the elements into order. In the modern periodic table the elements are in order of their atomic numbers. Use a modern periodic table to help you to answer these questions.

  (a) Give one example of a pair of elements that would be incorrectly placed if they were in order of their relative atomic masses. (1 mark)

  (b) How is the atomic number of an element related to the number of electrons in its atoms? (1 mark)

  (c) Explain the link between the electrons in an atom and its group in the periodic table. (1 mark)

  (d) Explain why using relative atomic masses does not put all elements into their correct groups. (2 marks)

  (e) Why were Newlands and Mendeleev unable to use atomic numbers? (1 mark)

2 Analysis of a sample of rainwater showed that only the following substances were dissolved in it:

  carbon dioxide, nitrogen, noble gases, oxygen, sulfur dioxide

  (a) Some of the rainwater was gradually heated to 80°C. Small bubbles appeared as the water was heated. Explain why. (2 marks)

  (b) The rainwater lathers easily with soap. Explain why. (2 marks)

  (c) This rainwater has a pH of 4.2. Which substances in the rainwater produce this low pH value? (1 mark)

  (d) A small amount of calcium hydroxide was added to the water. Suggest *two* ways that this would affect the water. (2 marks)

  (e) This water should be treated before it could be used for drinking water.

   (i) What method could be used to remove any solids? (1 mark)

   (ii) Suggest one method that could be used to make it safe to drink and explain how it works. (2 marks)

3 The graph shows the solubility curve for potassium nitrate.

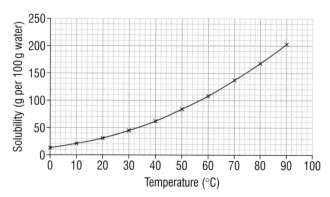

Solubility curve for potassium nitrate

  (a) Describe how the solubility of potassium nitrate changes with temperature. (1 mark)

  (b) What is the solubility of potassium nitrate at 65°C? (1 mark)

  (c) What mass of potassium nitrate would separate out if a saturated solution at 65°C were cooled to 20°C? (2 marks)

4 Some students wanted to find out how much citric acid was in this lemonade:

They put 150 cm³ of lemonade into a beaker and boiled it for 2 minutes to remove the carbon dioxide. They cooled the lemonade to room temperature. Then they titrated 25.0 cm³ samples with sodium hydroxide solution, concentration 0.200 moles per dm³. Their results are shown in the table.

Titration number	1	2	3
Volume of sodium hydroxide solution (cm³)	22.9	22.5	22.4

  (a) Why did they remove the carbon dioxide from the lemonade? (2 marks)

  (b) (i) What effect would boiling the lemonade have on the volume of their sample? (1 mark)

   (ii) What sort of error would this produce in their results? (1 mark)

  (c) The students calculated an average value of 22.45 cm³.

   (i) Which results did they use to do this? (1 mark)

   (ii) Suggest why they chose these results. (1 mark)

  (d) (i) Name an indicator that the students could use for this titration. (1 mark)

   (ii) Explain why you chose this indicator. (1 mark)

  (e) The equation for the reaction of citric acid with sodium hydroxide can be represented as shown:

   $H_3Cit + 3NaOH \rightarrow Na_3Cit + 3H_2O$

   (i) Calculate the number of moles of sodium hydroxide in 22.45 cm³ of solution. (2 marks)

   (ii) Calculate the number of moles of citric acid in 25.0 cm³ of lemonade. (1 mark)

   (iii) Calculate the concentration of citric acid in the lemonade in moles per dm³. (1 mark)

   (iv) Calculate the concentration of citric acid in grams per dm³. (The relative formula mass, M_r, of citric acid is 192.) (1 mark)

   [Higher]

 Test & Assessment Interactive quizzes, answers and hints online!

The answer is worth 6 out of the 8 marks.

The responses worth a mark are underlined in red.

We can improve the answer in several ways:

Francium is at the bottom of the group and it is not clear how Mendeleev knew that there was a gap. An extra mark could be gained by explaining that there were other known elements either side of the gap with masses greater and smaller than the predicted mass of francium, so he knew that an element was missing.

Francium (Fr) is a very rare element that was discovered in 1939. Mendeleev predicted the existence of francium in 1870.

(a) Explain why Mendeleev could predict that francium existed in 1870.
(3 marks)

(b) How would you expect francium to react with water? Compare the reaction with that of sodium with water and explain the reasons for your predictions.
(5 marks)

(a) Mendeleev put the elements into groups according to their properties and <u>left spaces for undiscovered elements.</u> He knew that there should be another element after caesium in Group 1 because <u>there was a gap there.</u>

(b) Francium would react with water <u>very violently and would explode.</u> It would produce <u>francium hydroxide,</u> which would be alkaline. It would react <u>much more vigorously than sodium.</u> This is because <u>its outer electron is further from the nucleus than the outer electron of sodium and so it would be more easily lost.</u>

The answer could be improved by mentioning that sodium and francium are both in Group 1, by making it clear that both elements have one electron in the highest occupied energy level, that they both form ions with a single positive charge, and that both elements produce hydrogen when they react with water.

The answer is worth 4 out of the 6 marks.

The responses worth a mark are underlined in red.

We can improve the answer in several ways:

A mark is given for a correct description of a strong acid, but the answer needs further explanation. It should explain that a weak acid does not ionise completely, so that there is a comparison of the two acids. An alternative would be to explain that the strong acid is completely ionised so it cannot ionise further, and diluting it only changes its concentration.

Student A described dilute hydrochloric acid as a weak acid. Student B said that it was still a strong acid even though it had been diluted.

(a) Which student was right? Explain your answer. *(2 marks)*

(b) Describe a test you could use to show the difference between a weak acid and a strong acid. Give the results you would expect for each acid.
(4 marks)

(a) Student B because a <u>strong acid is one that is completely ionised in water.</u>

(b) Add <u>magnesium ribbon</u> to solutions of both acids. The strong acid will <u>bubble vigorously giving off hydrogen</u> that pops with a burning splint. The weak acid will only <u>bubble slowly</u> and will not give enough hydrogen to get a pop.

The test for hydrogen and its result is not relevant here and should be omitted.

The important point that the concentrations of the two solutions must be the same has been missed.

Checklist

This spider diagram shows the topics in the unit. You can copy it out and add your notes and questions around it, or cross off each section when you feel confident you know it for your exams.

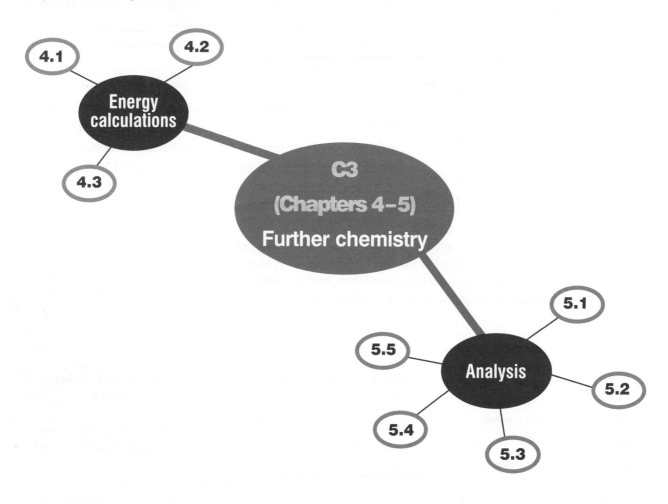

Tick when you:

reviewed it after your lesson	☑	☐	☐
revised once – some questions right	☑	☑	☐
revised twice – all questions right	☑	☑	☑

Move on to another topic when you have all three ticks.

Chapter 4 Energy calculations

4.1	Comparing the energy produced by fuels	☐	☐	☐
4.2	Energy changes in reactions	☐	☐	☐
4.3	Calculations using bond energies	☐	☐	☐

Chapter 5 Analysis

5.1	Tests for positive ions	☐	☐	☐
5.2	Tests for negative ions	☐	☐	☐
5.3	Testing for organic substances	☐	☐	☐
5.4	Instrumental analysis 1	☐	☐	☐
5.5	Instrumental analysis 2	☐	☐	☐

What are you expected to know?

Chapter 4 Energy calculations (See students' book pages 248–255)

- The amount of energy produced by different foods and fuels can be compared by heating water in a simple calorimeter.

- Energy is measured in joules (J) or calories (cal).

- Carbohydrates, fats and oils in foods produce relatively large amounts of energy.

- Eating food that provides more energy than the body needs can cause obesity.

- Energy is needed to break chemical bonds and energy is released when bonds are made.

- Energy level diagrams show the energy changes of chemical reactions.

- In exothermic reactions the energy needed to break the bonds in the reactants is less than the energy released when the bonds form in the products.

- The energy change of reactions that happen in solutions can be found by measuring the temperature change of the solution in an insulated container.

- Bond energies can be used to calculate the energy changes of reactions. [Higher Tier only]

Chapter 5 Analysis (See students' book pages 258–269)

- The different colours produced in flame tests by compounds containing lithium, sodium, potassium, calcium and barium.

- Carbonates react with dilute acids to produce carbon dioxide, which turns lime water milky.

- The colour changes that happen when copper carbonate and zinc carbonate decompose on heating.

- How aluminium, calcium, magnesium, copper(II), iron(II) and iron(III) ions react with sodium hydroxide solution.

- The tests for nitrate, sulfate and halide ions.

- The tests for ammonium ions and ammonia gas.

- Organic compounds burn or char when heated in air.

- Unsaturated organic compounds decolourise bromine water.

- How to calculate the empirical formula of an organic compound from data about its combustion products. [Higher Tier only]

- Elements and compounds can be detected and identified by instrumental methods of analysis.

- Instrumental methods of analysis can be very accurate, sensitive and rapid.

- Atomic absorption spectroscopy and mass spectrometry can be used to identify elements. [Higher Tier only]

- Compounds can be identified using methods that include infra-red spectrometry, ultraviolet spectroscopy, nuclear magnetic resonance spectroscopy and gas-liquid chromatography. [Higher Tier only]

1. What method could you use to compare the amount of heat produced when two substances burn?

2. Why is it difficult to obtain accurate results for the amount of heat produced when substances burn?

3. What happens to the bonds in the reactants and products during a chemical reaction?

4. What are the main features of an energy level diagram for a reaction?

5. What are bond energies?

6. What can we use bond energies for? [Higher Tier only]

---

students' book
page 248

C3 4.1   Comparing the energy produced by fuels

KEY POINTS

1 Heat energy is produced when fuels burn.
2 We can use a simple calorimeter to compare the energy produced when different substances burn.

AQA EXAMINER SAYS...

Energy values that have been accurately measured can be found in tables of data. These values can be used to make more reliable comparisons of different fuels and foods. You may be given this type of information in examination questions.

When different substances burn and different foods are eaten, different amounts of energy are produced. Eating food that gives you more energy than your body needs can lead to obesity.

Energy is measured in joules (J). Values for heat energy and dietary information are sometimes given in calories (1 cal = 4.2 J).

We can use a calorimeter to measure the amount of heat produced when substances burn. Accurate energy measurements need carefully controlled conditions and special apparatus.

Simple calorimeters can be used to compare the amount of heat produced by different fuels.

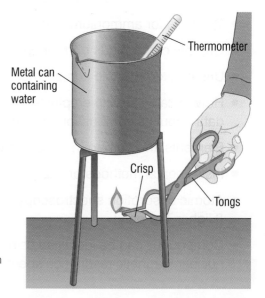

Thermometer

Metal can containing water

Crisp

Tongs

The energy released by foods when they burn can be compared using some very simple equipment

Results are not accurate when using a simple calorimeter because it is difficult to control the variables.

Remember: 1 kJ = 1 kilojoule = 1000 joules

The simplest calorimeter is some water in a glass beaker or metal can. When a substance burns and heats the water, the temperature rise of the water depends on the amount of heat produced:

- 4.2 J of heat energy increases the temperature of 1 g of water by 1°C.

Simple calorimeters do not give accurate results for the energy produced because much of it is used to heat the surroundings.

Key words: joule, calorie, calorimeter

CHECK YOURSELF

1 What is the apparatus called that we use to measure heat energy?

2 What unit is used to measure energy?

3 Why is special apparatus needed to measure energy values accurately?

4 What measurements would you make when using a simple calorimeter to compare the heat produced by burning two substances?

---

students' book page 250

C3 4.2 Energy changes in reactions

KEY POINTS

1 Energy is used to break chemical bonds and is released when bonds are made.

2 In an exothermic reaction, the energy released by making bonds in the products is greater than the energy needed to break the bonds in the reactants.

EXAMINER SAYS...

You should be able to interpret energy level diagrams in terms of bond breaking and bond making, activation energy and energy change of the reaction.

EXAM HINTS

Make sure you can draw and label energy level diagrams for exothermic and endothermic reactions.

In a chemical reaction new substances are formed. The bonds in the reactants must be broken for reactions to happen:

- The energy needed to break these bonds is the activation energy for the reaction.

When new bonds form to make the products energy is transferred to the surroundings:

- If more energy is needed to break bonds than is released when new bonds form, the chemical reaction is endothermic.
- If more energy is released when the bonds are made in the products than is used in breaking the bonds in the reactants, the chemical reaction is exothermic.

These changes can be represented on an energy level diagram. The difference between the energy levels of reactants and products is the energy change for the reaction. This energy change is represented by the symbol ΔH:

- ΔH is negative for an exothermic reaction.
- ΔH is positive for an endothermic reaction.

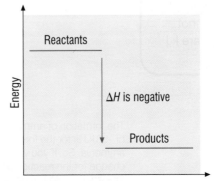

An exothermic reaction – more energy is released when bonds are formed between atoms in the products than is needed to break the bonds between the atoms in the reactants

GET IT RIGHT!

Energy level diagrams show the relative energies of reactants and products.

EXAM HINTS

For exothermic reactions the products are lower on the energy level diagram than the reactants.

Energy changes for reactions are measured in kJ per mole. The energy change of reactions in solutions can be measured using an insulated plastic container.

Key words: activation energy, energy level diagram

CHECK YOURSELF

1 What must happen to the reactants for a reaction to start?

2 What makes a reaction exothermic?

3 How is the energy change for a reaction shown on an energy level diagram?

4 How could you measure the energy change of the reaction between ammonia solution and dilute nitric acid?

HIGHER

students' book
page 252

C3 4.3 **Calculations using bond energies**

KEY POINTS

1 The amount of energy needed to break a chemical bond is called its 'bond energy'.

2 Energy changes for reactions can be calculated using bond energies.

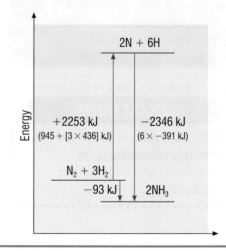

AQA EXAMINER SAYS...

Calculations of this type will appear only on Higher Tier papers.

BUMP UP YOUR GRADE

Make sure you use the mole quantities in the balanced equation in your calculation, and do not forget the units for energy are kJ for these quantities.

Energy is needed to break chemical bonds. Different amounts of energy are needed to break different bonds. The amount of energy needed to break a particular bond is called its 'bond energy'.

Bond energies are measured in kilojoules per mole (kJ/mol) and are listed in tables. Bond energies can be used to calculate the energy changes (ΔH) of reactions as follows:

1 Use the balanced equation for the reaction to find the number of each type of bond in the reactants. (It helps to draw out the structures of any molecules.)

2 Calculate the total amount of energy needed to break these bonds.

3 Use the balanced equation to find the number of each type of bond in the products.

4 Calculate the total amount of energy released when these bonds are made.

5 The difference between the two totals is the energy change for the reaction.

Key words: bond energy

The formation of ammonia. The energy change of –93 kJ is for the formation of *two* moles of ammonia. So, if you wanted to know the energy change for the reaction *per mole of ammonia* formed, it would be exactly half this, i.e. –46.5 kJ/mol.

GET IT RIGHT!

Bond breaking is endothermic (+), bond making is exothermic (–).

EXAM HINTS

You will be given the bond energies you need in the question and sometimes values for bonds that you do not need to use in your calculation, so check carefully.

CHECK YOURSELF

1 What are the units for bond energies?

2 What do you need to know about the reactants before you can calculate the total energy needed to break the bonds?

3 Calculate the energy change for the reaction $CH_4 + 2O_2 \rightarrow CO_2 + 2H_2O$ using bond energies in the table:

Bond	Bond energy (kJ/mol)	Bond	Bond energy (kJ/mol)
C–H	413	O=O	498
C=O	805	H–O	464

C3 4   End of chapter questions

1 What simple apparatus would you use to compare the energy produced by two different liquid fuels?

2 Why is the apparatus in question 1 not the best way to compare the fuels?

3 How do the energy changes for bond making and bond breaking differ for an exothermic reaction?

4 Draw and label an energy level diagram for an endothermic reaction.

5 Describe the measurements you would make to find the energy change for the reaction of calcium oxide with water using an insulated plastic cup.

6 Calculate the energy change for the reaction $H_2 + Br_2 \rightarrow 2HBr$ using bond energies: H–H = 436 kJ/mol, Br–Br = 193 kJ/mol, H–Br = 366 kJ/mol. [Higher Tier only]

(1) What are the two main tests you can use for metal ions?

(2) Which metal ions are coloured when in solution and form coloured hydroxides?

(3) What is the test for carbonate ions?

(4) What is the test for halide ions?

(5) What happens to organic compounds when they are heated in air?

(6) How can you find the empirical formula of an organic compound? [Higher Tier only]

(7) Name two instrumental methods that can be used to detect elements. [Higher Tier only]

(8) Which technological advances have helped the development of instrumental methods?

(9) What method uses a moving gas or liquid to separate compounds in a mixture? [Higher Tier only]

(10) Name two instrumental methods of analysis that give information about the structure of molecules. [Higher Tier only]

---

students' book page 258

C3 5.1   Tests for positive ions

KEY POINTS

1 Some metal ions produce colours in a flame.
2 Sodium hydroxide solution reacts with many positive ions in solution.

EXAMINER SAYS...

You should know how to do these tests in the laboratory. You should be aware that metal ions not included in the GCSE specification may give results in these tests, but you are not expected to know the results for these ions.

To identify unknown substances there are tests that can help us tell one substance from another. Many positive ions can be identified using a flame test or using sodium hydroxide solution.

Some metal ions produce colours when put into a flame:

Element	Flame colour
lithium	bright red
sodium	golden yellow
potassium	lilac
calcium	brick red
barium	green
magnesium	no colour

Most hydroxides of metals that have ions with 2+ and 3+ charges are insoluble in water. When sodium hydroxide is added to solutions of these ions a precipitate forms.

GET IT RIGHT!

Make sure you know the results for the 11 positive ions named in this section.

We can show these reactions in an ionic equation, e.g.

$$Al^{3+}(aq) + 3OH^-(aq) \rightarrow Al(OH)_3(s)$$

Key words: flame test

BUMP UP YOUR GRADE

Top grade candidates can write ionic equations for the reactions of positive ions with hydroxide ions.

- Aluminium, calcium and magnesium ions form white precipitates. Aluminium hydroxide dissolves when excess sodium hydroxide solution is added.
- Copper(II) hydroxide is blue.
- Iron(II) hydroxide is green and slowly turns brown.
- Iron(III) hydroxide is reddish brown.

Ammonium ions react with sodium hydroxide solution to form ammonia. If the solution is warmed ammonia gas is given off and turns damp red litmus paper blue.

CHECK YOURSELF

1 List the ions named in this section that give colours in a flame test.
2 List the ions that give white precipitates with sodium hydroxide.
3 List the positive ions that give coloured precipitates with sodium hydroxide solution.
4 What is the test and result for an ammonium ion?
5 Write an ionic equation for one of the reactions in question 3. [Higher Tier only]

---

students' book page 260

C3 5.2 Tests for negative ions

KEY POINTS

1 Carbonates produce carbon dioxide when they are heated or reacted with acids.
2 Halide and sulfate ions can be detected by precipitation reactions.
3 Nitrate ions can be reduced to form ammonia.

There are four main tests for negative ions:

- **Carbonate ions:** Add dilute hydrochloric acid to the substance to see if it fizzes. If it does and the gas produced turns lime water milky, the substance contains carbonate ions.

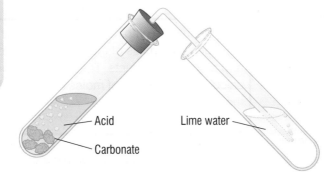

Acid

Carbonate

Lime water

The test for carbonates

- **Halide ions:** Add dilute nitric acid and then silver nitrate solution. Chloride ions give a white precipitate, bromide ions a cream precipitate and iodide ions a yellow precipitate.
- **Sulfate ions:** Add dilute hydrochloric acid and then barium chloride solution. If a white precipitate forms, sulfate ions are present.
- **Nitrate ions:** Add a little aluminium powder and then add sodium hydroxide solution. Gently warm and test the gas given off with damp red litmus paper. If it turns blue, ammonia was produced and nitrate ions are present.

Carbonates also produce carbon dioxide when they decompose on heating but this can be more difficult to detect. Copper carbonate is green and when heated it decomposes to black copper oxide. Zinc carbonate is white and decomposes to form zinc oxide, which is yellow when hot but turns white when cool.

Key words: ion, precipitate

BUMP UP YOUR GRADE

Top grade candidates can apply their knowledge of these tests to solving problems, so practise answering this type of question.

GET IT RIGHT!

Add <u>nitric</u> acid with silver <u>nitrate</u>, hydro<u>chloric</u> acid with barium <u>chloride</u>.

CHECK YOURSELF

1 Which four negative ions in this section are tested using precipitation reactions?

2 When testing for carbonate ions, why is the gas tested with lime water?

3 Why should you check that there are no ammonium ions present before testing for nitrate ions?

4 What is unusual about zinc carbonate?

EXAM HINTS

Barium sulfate, silver halides and carbonates of metals with 2+ and 3+ ions are insoluble in water.

KEY POINTS

1 Organic compounds burn or char when heated in air.

2 The empirical formula of an organic compound can be found from the masses of reactants and products of combustion reactions. [Higher Tier only]

3 Unsaturated organic compounds react with bromine and iodine.

All organic compounds contain carbon. Many organic compounds burn when heated in air and produce carbon dioxide. Those that do not burn usually blacken or char showing that they contain carbon.

We can find the empirical formula of organic compounds that burn by measuring the mass of the compound and the masses of carbon dioxide, water and any other products. The masses of the products are converted into moles to calculate the number of moles of carbon, hydrogen and any other elements in the compound.

Organic compounds that contain a carbon–carbon double bond are called 'unsaturated' and react with bromine and iodine:

● Bromine water is decolourised in these reactions – it is used to test whether or not substances are unsaturated.

● Iodine solution is used in titrations to find the number of double bonds in organic molecules such as fats and oils.

Key words: char, empirical formula, unsaturated

HIGHER

 EXAMINER SAYS...

Higher Tier questions on this section will often require you to do calculations and to solve problems. Make sure that you have revised your work on the mole and acid–alkali titrations.

Worked example

An organic substance Z contains carbon and hydrogen. A sample of Z is burnt in an excess of oxygen, producing 1.80 g of water and 3.52 g of carbon dioxide. What is the empirical formula of Z?

Solution

Step 1: Calculate moles of CO_2:

The M_r of CO_2 is $12 + (2 \times 16) = 44\,g$

Amount of $CO_2 = \dfrac{3.52}{44} = 0.08\,moles$

Step 2: Calculate moles of H_2O:

The A_r of H_2O is $(2 \times 1) + 16 = 18\,g$

Amount of $H_2O = \dfrac{1.80}{18} = 0.10\,moles$

Each molecule of carbon dioxide formed requires one carbon atom from a molecule of Z. So for each mole of carbon dioxide formed, Z must contain one mole of carbon atoms.

Amount of C atoms in sample of $Z = 0.08$ moles

In the same way, each water molecule formed requires two hydrogen atoms from a molecule of Z. So for each mole of water formed, Z must contain two moles of hydrogen atoms.

Amount of H atoms in sample of $Z = 0.10 \times 2 = 0.20$ moles

So Z contains carbon atoms and hydrogen atoms in the ratio $0.08 : 0.20 = 2 : 5$.

Therefore the empirical formula of Z is C_2H_5.

CHECK YOURSELF

1 Why do some compounds char when heated?

2 What information would you need to calculate the empirical formula of an organic compound that burns in air? [Higher Tier only]

3 Why can bromine be used as a test for unsaturation in organic compounds?

4 What apparatus and solutions would you need to do a titration to find the amount of iodine that reacted with an unsaturated oil?

C3 5.4 Instrumental analysis 1

KEY POINTS

1 Instrumental methods used to detect and identify elements and compounds are accurate, sensitive and rapid.

2 Elements can be detected by atomic absorption spectroscopy and mass spectrometry. [Higher Tier only]

Identification and measurement of elements and compounds is important in industry, in health care and in monitoring the environment. Instrumental methods of analysis are rapid, accurate and sensitive, often using very small samples. Machines can carry out an analysis very quickly and accurately. Computers calculate and process the data from the instrument to give meaningful results almost instantly.

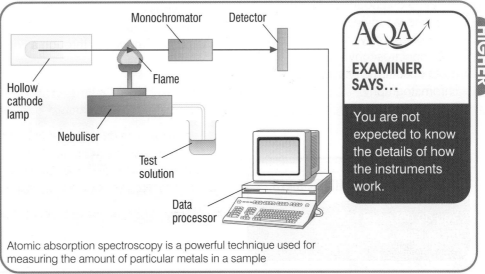

Atomic absorption spectroscopy is a powerful technique used for measuring the amount of particular metals in a sample

AQA

EXAMINER SAYS...

You are not expected to know the details of how the instruments work.

Electric and magnetic fields ensure all ions have the same kinetic energy

Sample injected and ionised

Recording equipment

Magnetic field deflects ions

Beam of ions

A mass spectrometer provides an accurate way of measuring the mass of atoms

Two methods that are used for detecting and measuring elements in samples are:

- **Atomic absorption spectroscopy**, this can detect very small amounts of about 40 elements, mainly metals, in samples as small as $0.02\,cm^3$.
- **Mass spectrometers**, these can be used to identify and measure the elements in a substance. Mass spectrometry can find the ratios of isotopes of an element that can be used to calculate relative atomic masses.

Key words: atomic absorption spectroscopy, mass spectrometer

BUMP UP YOUR GRADE

Make sure you know one or two examples of how each method is used.

GET IT RIGHT!

Computers are used to automate instruments and to process data rapidly.

CHECK YOURSELF

1 Why are instrumental methods often preferred to traditional laboratory methods of chemical analysis?

2 How have developments in electronics and computers helped in the development of instrumental methods?

3 What is atomic absorption spectroscopy used for? [Higher Tier only]

4 How is mass spectrometry used? [Higher Tier only]

KEY POINT

Methods used to identify and measure compounds include
IR spectroscopy,
UV spectroscopy,
NMR spectroscopy,
mass spectrometry and chromatography.

Instrumental methods are often automated, more sophisticated versions of methods we use in the laboratory. Chromatography separates compounds in a mixture, according to how well they dissolve in a certain type of solvent. Several different types of chromatography have been developed to separate and identify particular compounds. These include gas–liquid chromatography that uses a gas to carry compounds through a liquid on a solid support.

AQA EXAMINER SAYS...

You do not need to know how these instruments work.

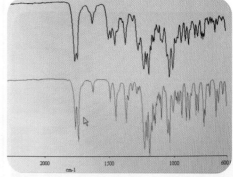

An infra-red spectrometer

HIGHER

EXAM HINTS

It will help you to answer
questions on this topic if you have
seen photographs or video clips of
some of the instruments used.

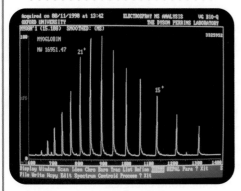

Data from a mass spectrometer

GET IT RIGHT!

Spectroscopic methods depend
on different substances absorbing
different types of radiation.

Spectroscopic methods use radiation. How this is absorbed by a sample
depends on the atomic and molecular structures of the substances it contains:

- Infra-red spectroscopy gives information about particular bonds in a
molecule.
- Nuclear magnetic resonance spectroscopy is used to find the structures of
organic molecules.
- Ultraviolet and visible light spectroscopy is useful for finding the amount of a
substance in a solution.
- Molecules break up into fragments in a mass spectrometer, which can give
information about their structures.

Key words: chromatography, spectroscopy

CHECK YOURSELF

1 What is chromatography used for?

2 What is passed through the sample in spectroscopic analysis?

3 Name three spectroscopic methods of analysis.

4 Which instrumental methods can give information about the structures of
molecules?

C3 5   End of chapter questions

1 Which ions produce a white precipitate when sodium hydroxide is added to the
solution and give a red colour in a flame test?

2 Which positive ions react with sodium hydroxide to produce a gas that turns damp
red litmus blue?

3 Name the green compound that reacts with dilute hydrochloric acid to produce a gas
that turns lime water milky and when heated produces a black solid.

4 Which ions give a yellow precipitate when nitric acid and silver nitrate solution are
added to a solution?

5 What is the test for unsaturation in an organic molecule and what is the result for an
unsaturated compound?

6 A hydrocarbon was burned completely in air and produced 0.44 g of carbon dioxide
and 0.27 g of water. What is its empirical formula? [Higher Tier only]

7 How have computers helped with the development of instrumental methods of
chemical analysis?

8 Suggest three reasons why atomic absorption spectroscopy is especially useful in
analysing water for polluting metals.

9 Why is chromatography often used before mass spectrometry when analysing
mixtures of organic compounds? [Higher Tier only]

10 Which spectroscopic method gives information about the bonds in molecules?
[Higher Tier only]

1 A student ate a beefburger in a bun (250 g) with French fries (150 g) and a cola drink (500 g). The student's friend suggested a chocolate bar (120 g), low fat crisps (25 g) and orange juice (400 g) would have been better. The table gives some nutritional information.

Food	Energy (kJ per 100 g)	Fat (g per 100 g)
Beefburger in bun	956	11
Chocolate bar	2095	30
Cola	176	0
Low fat crisps	2116	26
French fries	1174	15
Orange juice	153	0.1

(a) Which food contains the most fat per 100 g? (1 mark)
(b) Which food produces the most energy per 100 g? (1 mark)
(c) Calculate the total energy produced by each of these meals. (4 marks)
(d) Calculate the total fat content for each of these meals. (2 marks)
(e) Explain one way in which the friend's suggestion could be better for the student. (2 marks)
(f) Suggest one reason why the friend's suggestion may not be good advice. (1 mark)

2 Some students put four colourless aqueous solutions into four bottles but did not label the bottles. The solutions were:
• ammonium sulfate solution, $(NH_4)_2SO_4(aq)$
• calcium chloride solution, $CaCl_2(aq)$
• magnesium chloride solution, $MgCl_2(aq)$
• potassium sulfate solution, $K_2SO_4(aq)$
Describe a series of tests that you could do to identify the solution in each bottle. (8 marks)

3 Flame emission spectroscopy is a method of analysis developed from flame tests. A very small sample of solution is put into a flame in the instrument. When excited by the flame each element produces a unique pattern of radiation. This pattern can be detected and used to identify elements. The amount of radiation produced can be used to find the amount of the element in a sample. Using this method the concentration of sodium ions in blood can be measured accurately in a few seconds.
(a) What is the colour produced by sodium ions in a flame? (1 mark)

(b) Name another element that gives a colour in a flame and state the colour. (2 marks)
(c) Why can this method be used to identify elements? (1 mark)
(d) Why can this method be used to find the concentration of an element in a solution? (1 mark)
(e) Suggest *two* reasons why flame emission spectroscopy is especially useful in hospitals. (2 marks)

4 Ethanol is a liquid used as a fuel.
(a) Describe how you would measure the heat energy produced by burning ethanol using a simple calorimeter. Your answer should include:
• a labelled diagram of the apparatus you would use
• the measurements you would make. (6 marks)
(b) The table shows some bond energies.

Bond	Bond energy (kJ/mol)	Bond	Bond energy (kJ/mol)
C–H	413	C=O	805
O=O	498	H–O	464
C–O	358	C–C	347

The equation for the complete combustion of ethanol is:

$$C_2H_5OH + 3O_2 \rightarrow 2CO_2 + 3H_2O$$

Ethanol has the structure:

$$\begin{array}{ccc} & H & H \\ & | & | \\ H - & C - C & - O - H \\ & | & | \\ & H & H \end{array}$$

Use the bond energies in the table to calculate:
(i) the energy to break all the bonds in the reactants (2 marks)
(ii) the energy produced when the bonds in the products are made (2 marks)
(iii) the energy released by the reaction. (1 mark)
[Higher]

5 An organic compound X contains carbon and hydrogen only.
(a) X decolourised aqueous bromine water. What does this tell you about the organic compound? (1 mark)
(b) 0.35 g of X were burnt completely and produced 1.10 g of carbon dioxide and 0.45 g of water. Use this information to calculate the empirical formula of X. (Use data sheet on page 114.) (3 marks)
(c) The relative mass of X is 56. What is the formula of a molecule of X? (1 mark)
[Higher]

 Test & Assessment Interactive quizzes, answers and hints online!

The answer is worth 4 out of the 6 marks in this Higher Tier question.

The responses worth a mark are underlined in red.

We can improve the answer in several ways:

A good start drawing the structures of the reactants and products to show the bonds that must be counted. Unfortunately, the bonds in carbon dioxide are double bonds (carbon forms 4 bonds), so a mark is lost.

One mark is gained for correct working. It is always worth showing your working.

The fourth mark is gained as the error was carried forward.

The equation represents the complete combustion of ethane in air:

$$2C_2H_6 + 7O_2 \rightarrow 4CO_2 + 6H_2O$$

Bond	Bond energy (kJ/mol)	Bond	Bond energy (kJ/mol)
C–H	413	C=O	805
O=O	498	H–O	464
C–O	358	C–C	347

Use the bond energies in the table to calculate the energy needed to break all the bonds in the reactants, the energy released when new bonds are formed in the products and the energy change for the reaction. Explain how you can tell from this information that the reaction is exothermic. *(6 marks)*

$$2 \; H-\overset{\overset{\displaystyle H}{|}}{\underset{\underset{\displaystyle H}{|}}{C}}-\overset{\overset{\displaystyle H}{|}}{\underset{\underset{\displaystyle H}{|}}{C}}-H + 7\,O=O \rightarrow 4\,O-C-O + 6\,H-O-H$$

Breaking = 12 × 413 + 2 × 347 + 7 × 498 = 9136 kJ

Making = 4 × 2 × 358 + 6 × 2 × 464 = 8432 kJ

Energy change = 704 kJ

Reaction is exothermic because heat is given out.

The final part is a correct statement but does not use the information in the question or the calculation. According to this answer the reaction would be endothermic, because more energy was needed to break the bonds than was released when bonds were made. This is clearly wrong and should have alerted the student that a mistake had been made.

The answer is worth 4 out of the 6 marks.

The responses worth a mark are underlined in red.

We can improve the answer in several ways:

The precipitate dissolves in excess sodium hydroxide solution. The student has only made two valid points and should have realised that there are three marks for this part.

A sample of tap water was thought to contain aluminium ions.
(a) Describe a chemical test for aluminium ions that you could do in a school laboratory and give the result of the test. *(3 marks)*
(b) Aluminium ions can be detected by atomic absorption spectroscopy. What are the advantages of instrumental methods compared with chemical tests? *(3 marks)*

(a) Add sodium hydroxide solution to the tap water. A white precipitate shows aluminium ions are present.

(b) Instrumental methods are faster than chemical tests and can detect very small amounts of ions. They are a lot better than chemical tests and so are preferred by industry.

This does not say why they are better, and although they are preferred by industry this is not an advantage. The answer could have included either that they are more accurate or that they only require small samples.

Chapter 1

Pre Test

1 A compound.
2 Their atoms join with other atoms by giving taking or sharing electrons.
3 As a building material, and to make quicklime, cement and glass. (It has several other uses, but these are the ones you need to know.)
4 Calcium oxide and carbon dioxide.
5 All metal carbonates decompose when heated strongly enough.
6 $MgCO_3 \rightarrow MgO + CO_2$
7 By adding water to quicklime.
8 Calcium carbonate is formed.
9 By mixing cement, sand, stones or crushed rock and water.
10 To allow light in and to make them weatherproof.

Check yourself

1.1

1 Elements.
2 Atoms (of elements).
3 The atoms of two or more elements combined or bonded together.

1.2

1 Building material, to make quicklime, to make cement and to make glass.
2 Breaking down using heat.
3 Three.
4 Five.

1.3

1 Zinc oxide and carbon dioxide.
2 Sodium carbonate and potassium carbonate.
3 $CuCO_3 \rightarrow CuO + CO_2$
4 10.0 g

1.4

1 carbon dioxide + calcium hydroxide → calcium carbonate + water
2 $CO_2 + Ca(OH)_2 \rightarrow CaCO_3 + H_2O$
3 It produces calcium carbonate which is insoluble in water.
4 Calcium hydroxide reacts with carbon dioxide (from the air).

1.5

1 It is stronger, sets faster and sets under water.
2 A mixture of cement, sand, stones or crushed rock and water.
3 By reinforcing with steel.
4 It allows light through (transparent) and is weatherproof (insulator / does not corrode).

End of chapter questions

1 CO_2 and $MgCO_3$.
2 By giving taking or sharing electrons.
3 As a building material, and to make quicklime, cement and glass.
4 calcium carbonate → calcium oxide + carbon dioxide
5 Atoms are not created or destroyed in chemical reactions.
6 Sodium carbonate or potassium carbonate.
7 $Ca(OH)_2 + CO_2 \rightarrow CaCO_3 + H_2O$
8 By mixing slaked lime, sand and water.
9 By heating a mixture of limestone and clay.
10 By changing the proportions of the raw materials, and by adding other substances.

Chapter 2

Pre Test

1 Rock that contains enough of a metal or its compounds to make it economic to extract the metal.
2 Displacement, reduction with carbon and electrolysis.
3 Reduction with carbon in a blast furnace.
4 It is iron from the blast furnace that is about 96% iron.
5 It has a regular structure with layers that can slide over each other.
6 Carbon steels, low-alloy steels, high-alloy steels.
7 Metals that contain other elements.
8 Alloys with special properties, such as returning to shape after being bent.
9 Good conductors of heat and electricity, strong, hard and dense, can be bent or hammered into shape, high melting points (except for mercury).
10 By smelting and electrolysis.
11 Low density and resistant to corrosion.
12 High energy costs because of high temperatures and electrolysis.

Check yourself

2.1

1 Rocks that contain enough metal to make it worth extracting.
2 A metal that is found as the metal (not as a compound).
3 Any two metals from zinc, iron, tin, lead, copper (any two in the reactivity series below carbon and above silver).
4 One of: aluminium, magnesium, calcium, sodium or potassium (any metal above carbon in the reactivity series).
5 Oxygen is removed (from a compound).

2.2

1 Iron(III) oxide.
2 Coke.
3 iron oxide + carbon → iron + carbon dioxide
(OR iron oxide + carbon monoxide → iron + carbon dioxide)
4 It is brittle.

2.3

1 Its atoms are in regular layers that can slide over each other.
2 Alloys (mixtures) of iron and other elements.
3 1.5%
4 Low-alloy steel.

2.4

1 Hardness, strength, appearance and resistance to corrosion.
2 Because alloys are mixtures (not compounds).
3 They can return to their original shape (shape memory alloys).

2.5

1 In the central block.
2 They are all metals, good conductors of heat and electricity, strong, hard and dense, can be bent or hammered into shape, have high melting points (except mercury).
3 High-grade ores are running out, existing methods involve moving huge amounts of rock and use large amounts of energy.

2.6

1 Low density and resistance to corrosion.
2 Because it reacts with carbon (which makes it brittle).
3 It is too reactive to extract with carbon.

4 Resources (ores) and energy costs (of extracting the metal) are saved.

End of chapter questions

1 Enough to make it worth extracting.
2 Heat with carbon (displace / heat with a more reactive metal or electrolysis could be used but would be more expensive than using carbon).
3 Oxygen is removed (from iron oxide).
4 It contains (4%) impurities.
5 Iron and carbon.
6 Contains more than 5% of other metals.
7 Changes their properties / makes them more useful.
8 Smart OR shape memory alloys.
9 One of: strong, hard, dense, can be hammered into shape, high melting points.
10 One of: can extract low-grade ores, uses less energy, less rock to be moved / mined (but not just 'cheaper' – too vague).
11 Oxide layer.
12 To extract the (reactive) metal (sodium / magnesium), used to displace titanium.

Chapter 3

Pre Test

1 Hydrogen and carbon.
2 Compounds / (saturated) hydrocarbons with the formula C_nH_{2n+2}.
3 Fractional distillation.
4 Mixtures of compounds with similar boiling points separated from crude oil.
5 Carbon dioxide and water.
6 Carbon monoxide, carbon and unburnt hydrocarbons (particulates), sulfur dioxide.
7 Global warming, global dimming, acid rain.
8 Remove sulfur from fuels, remove sulfur dioxide from waste gases, use alternative fuels.

Check yourself

3.1

1 It is not useful as a fuel / it contains too many substances / substances in it burn under different conditions.
2 Distillation.
3 Compounds of carbon and hydrogen **only**.
4 By molecular formulae and structural formulae.

3.2

1 Fractional distillation.
2 By evaporating the crude oil and condensing the vapours at different temperatures.
3 Hydrocarbons with smallest molecules / lowest boiling points.
4 Those with high boiling points / large molecules / from near the bottom of the column.

3.3

1 Carbon dioxide and water.
2 Carbon monoxide, carbon and unburnt hydrocarbons / particulates.
3 Fuels contain compounds of sulfur.

3.4

1 Carbon dioxide is produced and it is a greenhouse gas.
2 Particles in the atmosphere reflect sunlight away from the Earth.
3 By removing sulfur compounds from fuels at the refinery, or by removing sulfur dioxide from the waste gases after burning.

End of chapter questions

1 It is a mixture (of liquids).
2 C_4H_{10}.
3 Evaporation and condensation.
4 Near the top of the column / tower.
5 propane + oxygen → carbon dioxide + water
6 Carbon and unburnt hydrocarbons.
7 Produces carbon dioxide (which is a greenhouse gas).
8 Sulfur.

EXAMINATION-STYLE QUESTIONS

1 (a) Gold and silver. (1 mark)
  (b) 0.3% (1 mark)
  (c) It was not worthwhile extracting the copper / they did not have methods to get the copper. (1 mark)
  (d) Need to move / dig up large amounts of rock.
    To get enough copper / to make it economical / because the ore is low grade. (2 marks)
  (e) (Up to 5 marks but max. 3 for either advantages or disadvantages)

Advantages:
● Produces a lot of copper.
● In a desert area.
● Area was in decline.
● New towns replaced old ones / investment in the area.
● Brought employment to the area.

Disadvantages:
● Huge pit is an eyesore / destroyed scenery.
● Dumping waste affects wildlife / landscapes.
● Brought in workers rather than local people. (5 marks)

2 (a) Iron(III) oxide. (1 mark)
  (b) Carbon (or coke). (1 mark)
  (c) Impurities / other elements / carbon, silicon, phosphorus. (1 mark)
  (d) Tool steel. (1 mark)
  (e) Stainless steel. (1 mark)
  (f) It has few impurities / it is almost pure iron / it has a regular structure. (1 mark)
    So the layers in the metal can slide over each other. (1 mark)

3 (a) It is a mixture (of liquids) with different boiling points. (1 mark)
  (b) To vaporise it / turn it into vapour or gas. (1 mark)
  (c) One mark each:
    ● The fractionating column is hot at the bottom and cooler at the top.
    ● The vapours move up the column and condense when they reach their boiling points.
    ● There are outlets at different levels to collect the liquid fractions.
    ● Hydrocarbons with the smallest molecules have the lowest boiling points and so are collected at the top of the tower / the fractions collected at the bottom have the highest boiling points. (4 marks)
  (d) (i) C_3H_8 (1 mark)
    (ii) Petroleum gases. (1 mark)
  (e) (i) Hydrocarbon molecules are larger. So more air / oxygen is required for complete combustion OR they do not burn as easily. (2 marks)
    (ii) Global dimming. (1 mark)

4 (a) Elements.
  (b) Metals.
  (c) Mixtures.
  (d) Compounds. (4 marks)

Chapter 4

Pre Test

1 Breaking down by heat (thermal decomposition) of large hydrocarbon molecules to make smaller molecules.
2 (Unsaturated) hydrocarbons with a carbon–carbon double bond (C=C).
3 A substance with long chain (very large) molecules made by joining many small molecules (monomers).
4 By addition polymerisation using ethene as the monomer.
5 Softens whenever it is heated.
6 It sets hard (chemical bonds form between the polymer chains).
7 By changing the monomers, by changing the reaction conditions when they are made.
8 Two of: food packaging / drinks bottles / breathable waterproof fabrics / medical uses / hydrogels / shape memory polymers / smart polymers.

Check yourself

4.1
1 (Catalytic) thermal decomposition.
2 Hot catalyst.
3 Unsaturated hydrocarbons / hydrocarbons with a double bond / general formula C_nH_{2n}.
4 Use bromine water – turns colourless.

4.2
1 Small molecules that join together to make polymers.
2 Poly(ethene).
3 The monomers add together to form the polymer and no other product is formed.
4 As plastics, e.g. bags, bottles, containers, toys.

4.3
1 They have very long molecules that tangle together.
2 It softens when it is heated (every time).
3 Chemical (strong) bonds are formed between the polymer chains.
4 Thermosoftening.

4.4
1 By changing the monomers / by changing the reaction conditions.
2 They keep food in good condition (or a specific property such as waterproof / transparent / do not tear easily / lightweight).
3 They can make fabrics that are waterproof but that are also able to let gases through (breathable).
4 One from: hydrogel, smart polymer, shape memory polymer.

End of chapter questions

1 To make smaller, more useful molecules / to make alkenes / to make smaller alkanes (for use as fuels).
2 They have two fewer hydrogen atoms / they have a double bond / they are more reactive / they react with bromine water.
3 Many monomers add / join together to form the polymer with no other product.
4 Propene / butene / pentene / hexene, etc.
5 Weak intermolecular forces.
6 The strong chemical bonds cannot be easily broken so it does not soften when heated.
7 Two of: lighter weight / flexible or not brittle or do not break easily.
8 Hydrogels.

Chapter 5

Pre Test

1 Foods and fuels.
2 By pressing or by distillation.
3 It cooks faster in oil than in water, because the oil is at a higher temperature than boiling water.
4 It has been reacted with hydrogen to remove the carbon–carbon double bonds (C=C) and make it saturated so that it is solid at room temperature.
5 A mixture of tiny droplets of immiscible liquids suspended in each other.
6 To prevent the liquids in the emulsion from separating.
7 A substance added to food to preserve it or to improve colour, flavour or texture.
8 These show additives permitted in the European Union.
9 Vegetable oil that can be used as fuel for diesel engines.
10 Renewable and cleaner / less harmful to the environment (less pollution).

Check yourself

5.1
1 Pressing and distillation (with water added).
2 They produce a lot of energy.
3 An oil with molecules that contain carbon–carbon double bonds (so they contain fewer hydrogen atoms than the corresponding saturated molecule).
4 Bromine water.

5.2
1 The temperature of hot oil is higher than boiling water.
2 Taste, colour, texture and energy content.
3 Addition reaction.
4 To make them solid at room temperature.

5.3
1 By vigorously mixing (stirring, shaking or beating) two liquids that do not usually mix together.
2 They keep the droplets suspended because different parts of their molecules are attracted to each of the liquids.
3 Thicker and more opaque / not transparent.

5.4
1 To preserve it, to improve colour, flavour and texture.
2 To show they are permitted additives (in the EU).
3 They would be listed in the ingredients.
4 They can be chemically analysed (using chromatography, mass spectrometry etc.).

5.5
1 It is produced from plants / crops that can be planted again.
2 Waste vegetable oil can be modified to use as diesel fuel (instead of dumping it or just burning it).

It produces less pollution (no additional carbon dioxide, no sulfur dioxide), more biodegradable than fossil fuels

End of chapter questions

1 The plant material is pressed and water and impurities are removed.
2 Use bromine water (it turns from orange-yellow to colourless).
3 Some of the oil is absorbed by the food.
4 Hydrogenation (reaction / addition of hydrogen using a catalyst).
5 Liquids that do not mix / dissolve in each other / are immiscible.
6 Salad dressings (mayonnaise) ice cream, milk, sauce.
7 Preservatives.
8 Chromatography.
9 The carbon dioxide released on burning was removed by plants as they grew.
10 Biodiesel is biodegradable / breaks down faster / less toxic.

Chapter 6

Pre Test
1 Crust, mantle and core.
2 Between 5 km and 70 km thick.
3 The crust and upper part of the mantle (the lithosphere).
4 At the bottom of the oceans (mid-oceanic ridges).
5 Carbon dioxide.
6 By (green) plants.
7 Nitrogen.
8 Noble gases.
9 Respiration, decomposition and combustion.
10 Photosynthesis, dissolving.

Check yourself
6.1
1 0.005 to 0.07 cm (or 0.05 to 0.7 mm)
2 Outer core is liquid, inner core is solid.
3 They thought the crust solidified and then the Earth cooled further causing it to shrink and the crust to wrinkle.

6.2
1 Large pieces of the Earth's crust and upper mantle (lithosphere).
2 Radioactivity releases heat that causes convection currents in the mantle, which move the plates.
3 Mountains form, earthquakes and volcanoes.

4 He could not explain why continents move / he was not a geologist.

6.3
1 Volcanoes.
2 It condensed to form the oceans.
3 Photosynthesis.

6.4
1 About 200 million years.
2 Argon.
3 In sedimentary rocks and fossil fuels.
4 They are very unreactive.

6.5
1 It returns to the atmosphere when animals respire or plants and animals die and decompose.
2 Dissolving in the oceans.
3 Burning fossil fuels.

End of chapter questions
1 0.5%
2 It is under the crust, about 3000 km thick / goes almost halfway to the centre, almost entirely solid, but can flow very slowly.
3 As plates move, huge forces are produced that cause the crust to crumple / buckle / deform.
4 New evidence from observations on the ocean floor.
5 Three from: carbon dioxide, nitrogen, water vapour, methane, ammonia.
6 Plants absorbed carbon dioxide to make food. Animals ate plants and produced shells that formed sedimentary rocks. Some plant and animal remains produced fossil fuels.
7 Noble gases (helium, neon, argon, krypton, xenon), carbon dioxide and water vapour.
8 It is less dense than air and it is not flammable / does not burn.
9 By heating up carbonate rocks (that have moved underground).
10 Carbon dioxide dissolves in the water, and forms carbonates that precipitate / form sediments.

EXAMINATION-STYLE QUESTIONS
1 (a) Circle drawn about half radius of first circle. (1 mark)
  (b) Core at centre, mantle between core and crust, crust is outer circle.
    (1 or 2 correct = 1 mark, all correct = 2 marks)
  (c) The crust is very thin / line is about the right thickness. (1 mark)

2 (a) Only 10% of the energy to make the plastic is needed to reshape it. (1 mark)
  (b) Two from:
    ● There are many different types.
    ● They look / appear to be very similar.
    ● Often made of mixtures of plastics. (2 marks)
  (c) Two from:
    ● Only a small mass used / not worth effort for so little plastic.
    ● Difficult to separate / not worth separating.
    ● Difficult to remove traces of food / low grade plastic produced.
    ● Less energy involved in making plastics than glass. (2 marks)
3 (a) Corn oil. (1 mark)
  (b) Olive oil. (1 mark)
  (c) Sunflower oil. (1 mark)
  (d) Solid below 25°C / melting point is too high. (1 mark)
  (e) Melting point decreases as number of double bonds increases. (1 mark)
4 (a) Three from:
    ● Place small amount of solvent in beaker.
    ● Stand / support paper upright in beaker OR place bottom of paper into solvent.
    ● Cover beaker with lid.
    ● Allow solvent to rise to top of paper. (3 marks)
  (b) To show where spots began / origin of spots / to show where to put the spots. (1 mark)
  (c) Ink would run / move up paper. (1 mark)
  (d) E124 and E127. (1 mark)
    Another unknown colour. (1 mark)
    Position of the spots / spots from sweet in line with these on paper. (1 mark)

C2 — Answers to questions

Chapter 1

Pre Test
1 Protons and neutrons in the nucleus with electrons surrounding it.
2 The number of protons in an atom of the element.
3 In energy levels (shells).
4 They have the same number of electrons in their highest energy level (outer shell).
5 They are stable arrangements.
6 They transfer (from metals to non-metals) or they are shared.
7 Strong electrostatic forces between oppositely charged ions.
8 It contains equal numbers of sodium and

chloride ions / the ratio of sodium ions to chloride ions is 1:1.
9 A pair of electrons shared by two atoms.
10 Either H–O–H or dot/cross diagram with a pair of dot/crosses between each of 2 Hs and an O and 4 more dot/crosses giving a total of 8 around the O.
11 In regular (giant) structures / lattices.
12 By metallic bonds – strong electrostatic forces between delocalised electrons and positive metal ions.

Check yourself
1.1
1 Protons, neutrons and electrons.
2 Proton +1, neutron 0, electron −1.
3 12

1.2
1 They represent electron energy levels.
2 Two concentric circles with 2 dots or crosses on the inner circle and 4 dots or crosses on the outer circle (and a dot at the centre for the nucleus).
3 Lithium 2,1; nitrogen 2,5; magnesium 2,8,2.

1.3
1 They have stable electronic arrangements/structures (not full outer shells).
2 K^+ 2,8,8; Mg^{2+} 2,8; O^{2-} 2,8.
3 Lithium atom: inner circle with 2 dots or crosses, outer circle with 1 dot or cross.
Fluorine atom: inner circle with 2 dots/crosses, outer circle with 7 dots/crosses.

Lithium ion: one circle with 2 dots/ crosses surrounded by square bracket with + sign outside top right of bracket.
Fluoride ion: as fluorine atom but with 8 dots/crosses in outer circle and square bracket with – sign outside top right.

1.4
1 Strong electrostatic forces between oppositely charged ions.
2 Equal numbers of sodium and chloride ions / ratio of 1 sodium ion : 1 chloride ion (no mention of molecules).
3 CaF_2

1.5
1 A shared pair of electrons (between two atoms).
2 4
3 N at centre with 3 separate Hs around it, with a dot and a cross between each H and the central N, and 2 more dots/crosses at a similar distance from the N (total 5 dots + 3 crosses or 3 dots + 5 crosses).

1.6
1 In regular patterns / lattices / giant structures.
2 On surface of some metals / when metals are displaced from solutions.
3 Electrons from highest energy level (outer electrons) that move freely within a metal (structure) or from atom to atom in a metal.
4 Electrostatic forces (between electrons and positive ions).

End of chapter questions
1 Both protons and electrons have the same charge and atoms are neutral.
2 9 protons, 9 electrons.
3 2,8,8,1.
4 All have one electron in their outer shell.
5 An unreactive gas, with a stable arrangement of electrons, in Group 0 of the periodic table.
6 The sodium atom loses its outer electron to form an ion with a single positive charge. The fluorine atom gains an electron to form an ion with a single negative charge. The sodium electron has transferred to the fluorine atom.
7 Strong electrostatic forces between oppositely charged ions (ionic bonds).
8 It has equal numbers of potassium ions and chloride ions (in the lattice).
9 The number of bonds formed = 8 – the Group number (8 minus the Group no.).
10 H–S–H is the simplest way to show this. Could also be shown as a dot/cross diagram with S and 2 Hs with a dot and cross between each of them, and another 4 dots or crosses around the S.
11 Close together in regular patterns / in layers / in a lattice.
12 By electrostatic forces between the delocalised (free) electrons and positive metal ions.

Chapter 2

Pre Test
1 They have giant structures with strong forces holding the ions together.
2 When they are molten or in solution.
3 There are only weak forces between the molecules so they melt and boil at low temperatures.
4 Molecules do not have a charge.
5 Their atoms can form several strong bonds with each other and/or other atoms.
6 They have very different structures: diamond is a giant 3-D structure, while graphite has giant 2-D layers.
7 The atoms are in layers that slide over each other into new positions.
8 Yes, all metals conduct heat and electricity.

9 The study of very small particles containing a few hundred atoms and about a few nanometres in size.
10 As sensors, catalysts, coatings and construction materials, and medical uses.

Check yourself
2.1
1 Many strong forces hold the ions together.
2 Water molecules can split up the lattice.
3 The ions are charged and can move (so they carry the electricity).

2.2
1 It is made of small molecules.
2 It is made of molecules and the molecules are not charged.
3 Weak intermolecular forces / weak attractions between molecules.

2.3
1 A giant covalent structure / a giant structure with atoms joined by covalent bonds.
2 They have regular three-dimensional giant structures.
3 It is made of flat (2-dimensional) molecules with weak forces between the layers.
4 It has delocalised electrons.

2.4
1 The layers of atoms can slide over each other into new positions (without the structure breaking apart).
2 The delocalised electrons move through the metal.
3 Delocalised electrons carry the energy.

2.5
1 A few hundred (200–500).
2 Their structure (may be different) and their very small size.
3 Three from: sensors, catalysts, coatings, construction materials, drug delivery/release, microprocessors/computers.

End of chapter questions
1 There are many strong forces to overcome to break down the giant structure so the ions can move about.
2 The ions are free to move in the solution.
3 Weak forces between molecules.
4 The molecules stay as molecules but they move apart and escape from the liquid. (The covalent bonds are not broken.)
5 Each atom in the giant structure is held by four strong covalent bonds, and many bonds have to be broken to free the atoms from the lattice.
6 They are covalently bonded in giant flat sheets.
7 The atoms are in layers that can slide over each other without breaking apart because the delocalised electrons keep them held together.
8 Only the delocalised electrons move through the metal (and any that leave the metal are replaced by other electrons from the electric current).
9 They are very small in size, have different structures, have very large surface areas, conduct more easily, and can be sensitive to light, heat, pH, electricity and magnetism.
10 The application of nanoparticles to specific uses.

Chapter 3

Pre Test
1 The number of protons plus the number of neutrons in the atom.
2 Isotopes
3 Atoms are too small to weigh so we use relative masses.
4 An amount of a substance, equal to its formula mass in grams, that contains a certain number of particles.

5 From its formula using relative atomic mass and relative formula mass.
6 Empirical formula is the simplest formula, molecular formula shows the number of atoms in a molecule.
7 The relative amounts of the substances in the reaction.
8 5.6 g
9 The amount of product actually made compared with the maximum amount it is possible to make.
10 It measures how much of the starting materials end up as useful product.
11 By its special symbol, and because it goes in both directions so products react to produce reactants.
12 When a reversible reaction is in a closed system and the rates of forward and backward reactions are equal.
13 By the Haber process from nitrogen and hydrogen.
14 It is a reversible reaction (and because of the conditions used).

Check yourself
3.1
1 Mass number.
2 Atoms of the same element with different numbers of neutrons / atoms with the same atomic number but different mass numbers.
3 $^{16}_{8}O$: p = 8, n = 8, e = 8; $^{19}_{9}F$: p = 9, n = 10, e = 9

3.2
1 (a) $H_2 = 2$
   (b) $CH_4 = 16$
   (c) $MgCl_2 = 95$
2 18 g
3 $^{12}_{6}C$

3.3
1 71% (or 71.4% or 71.43%)
2 CH_2
3 $FeCl_2$

3.4
1 $H_2 + Cl_2 \rightarrow 2HCl$
2 58.5 g
3 12.5 g

3.5
1 Reactions do not always go to completion and some product may be lost in the process.
2 69.8%
3 56%

3.6
1 A reaction that can go in both directions / a reaction in which products react to produce reactants.
2 It can reach equilibrium / rates of forward and backward reactions become equal.
3 By changing the conditions / by opening the system or allowing product to escape.

3.7
1 Nitrogen (air) and hydrogen (natural gas).
2 nitrogen + hydrogen ⇌ ammonia
   $N_2 + 3H_2 \rightleftharpoons 2NH_3$
3 It is reversible / ammonia breaks down into nitrogen and hydrogen.

End of chapter questions
1 Their mass is too small / negligible.
2 13 protons, 13 electrons, 14 neutrons.
3 62
4 44 g
5 75%
6 Fe_2O_3
7 $CH_4 + 2O_2 \rightarrow CO_2 + 2H_2O$
8 4.25 g
9 63%
10 74.4%
11 The reaction can go both ways: ammonium chloride reacts to produce ammonia and hydrogen chloride and hydrogen chloride and ammonia react to produce ammonium chloride.
12 A closed system.

13 450°C and 200 atmospheres pressure.

14 They are recycled.

1 (a) Three concentric circles with Na or dot at centre, 2 crosses or dots on inner circle, 8 on next and one on outer circle. (All correct = 2 marks; 1 mark for diagram with one error or omission.)

(b) Two concentric circles with Na or dot at centre, 2 crosses or dots on inner circle, 8 on outer circle (all enclosed in brackets), with + at top right side.
(All correct = 2 marks; 1 mark for diagram with one error or omission.)

(c) Circles with 2,8 as for part (b), with O or dot at centre, (all enclosed in brackets) with 2– at top right side. (All correct = 2 marks; 1 mark for diagram with one error or omission.)

(d) Giant lattice / giant structure / many ionic bonds to overcome. (1 mark) Ionic bonding is strong / oppositely charged ions strongly attract. (1 mark)

(e) Sodium ion has single (positive) charge and oxide ion has double (negative) charge, (1 mark) so lattice/structure/compound contains twice as many sodium ions as oxide ions *or* empirical formula has 2 Na to each O. (1 mark)

2 (a) (i) 1 (ii) 0 (iii) –1 (3 marks)

(b) (i) 17 protons, 18 neutrons, 17 electrons. (3 marks)

(ii) $^{37}_{17}Cl$ has 2 neutrons more than $^{35}_{17}Cl$. (1 mark)

(c) Both have the same number of electrons, (1 mark) so they have the same number in the outer shell / are in the same group of the periodic table / are the same element / chemical properties depend on the number of (outer) electrons. (1 mark)

3 (a) Covalent. (1 mark)

(b) N at centre with 8 dots or crosses around it, 3 H arranged around these, so that there is a pair of dots/crosses between each H and the central N.

e.g. H : N : H

H

(All correct = 2 marks; 1 mark for diagram with one error or omission.)

(c) Made of small molecules, (1 mark) with weak forces between the molecules. (1 mark)

4 (a) ammonium chloride $\rightleftharpoons$ ammonia + hydrogen chloride
(Correct names on either side of equation (1 mark); reversible reaction symbol (1 mark).)

(b) (i) 53.5 g (2 marks for correct answer with units, 1 mark for working: e.g. 14 + 4 + 35.5 or answer without g.)

(ii) 26.2% (2 marks for correct answer, accept 26.17%; 1 mark for working: e.g. 14 × 100 / 53.5.)

5 (a) germanium oxide + hydrogen $\rightarrow$ germanium + water (2 marks)
(1 mark for reactants, 1 mark for products; accept hydrogen oxide for water.)

(b) $GeO_2 + 4HCl \rightarrow GeCl_4 + 2H_2O$ (2 marks)
(1 mark for 4HCl, 1 mark for 2H₂O)

(c) Ge^{4+}, O^{2-} (2 marks)

(d) Central Ge surrounded by 4 pairs of dots/crosses, surrounded by 4 Cl, so that there is a pair of dots/crosses between each Cl and Ge.
(All correct = 2 marks; 1 mark for diagram with one error or omission.)

(e) (i) Two from:
Shiny (solid), conducts electricity, forms positive ions / ionic compound (with oxygen), oxide is base / reacts with acid to form a salt.
[High melting point is a neutral statement – could apply to both.] (2 marks)

(ii) Two from:
Brittle (solid), has a giant covalent structure, forms a covalent chloride / its chloride has small molecules, only conducts a small amount of electricity. (2 marks)

(f) (i) 62.4% (2 marks)
(1 mark for working: e.g. 73 × 100 / (73 + 32 + 12))

(ii) 67% (2 marks)
(1 mark for working: e.g. 73 × 100 / (73 + 36))

(iii) Two from:
Better atom economy, other product less harmful to environment / CO_2 produces global warming, water does not (or words to that effect), hydrogen better for sustainable development. (2 marks)

Chapter 4

Pre Test

1 How fast the reaction goes / how quickly reactants are used / how quickly products are formed.

2 By measuring the amount of a reactant used in a certain time or by measuring the amount of a product formed in a certain time.

3 They must collide with sufficient energy to react.

4 Changing conditions changes the frequency of collisions (the number of collisions in a given time, e.g. collisions per second).

5 Particles collide more frequently (more collisions per second) *and* with more energy.

6 The rate doubles.

7 It increases.

8 An increase in pressure increases the rate of reactions between gases.

9 A catalyst changes (increases) the rate of a reaction and is left at the end of the reaction.

10 They speed up reactions and reduce energy costs.

Check yourself

4.1

1 The rate of a reaction measures how fast the reaction goes.

2 Amount of reactant used or product formed and the time.

3 By measuring the volume or mass of gas produced over time (in a certain time).

4.2

1 Reactions happen when particles collide with enough energy to react.

2 Increasing temperature, concentration of solutions, pressure of gases, surface area of solids and using a catalyst.

3 Powders have more (greater) surface area than solids.

4.3

1 Both the number of collisions and the energy of the collisions are increased.

2 Increasing the temperature by 10°C.

3 Decreasing the temperature slows down the rate of reactions that spoil food.

4.4

1 Particles collide more frequently (more often).

2 The particles are closer together (concentration is increased) so they collide more frequently.

3 Equal volumes of the same concentration contain the same number of particles of solute.

4.5

1 They lower the activation energy so more collisions result in reactions.

2 They are not used up in the reaction / can be used over and over again.

3 Catalysts often work with only one type of reaction / are specific for a particular reaction.

End of chapter questions

1 rate = $\dfrac{\text{amount of reactant used}}{\text{time}}$

or $\dfrac{\text{amount of product formed}}{\text{time}}$

2 Measure volume of gas produced against time (accept mass lost against time *or* change in pH with time).

3 Activation energy.

4 Increase temperature, increase concentration of reactants, increase surface area of magnesium (use smaller pieces of magnesium), use a catalyst [*not increase pressure*].

5 Increasing temperature increases the frequency of collisions *and* the energy of the collisions.

6 It will be one quarter of the original rate.

7 The dissolved particles are closer together and so collide more frequently (more times per second).

8 The particles are closer at higher pressure so collide more often.

9 They lower the activation energy so more collisions result in reaction.

10 They are not used up (they are left at the end of the reaction).

Chapter 5

Pre Test

1 Exothermic.

2 Endothermic.

3 They are equal.

4 Increasing the temperature increases the amount of products (yield) from the endothermic reaction.

5 Those with different numbers of molecules of gases in the reactants compared to the number of molecules of gases in the products.

6 To produce ammonia as economically as possible.

Check yourself

5.1

1 Endothermic.

2 Heat is given out / the surroundings get hotter / temperature (of surroundings) increases.

3 When they dissolve (or react with saliva) endothermic reaction(s) happen.

5.2

1 Reverse reaction is exothermic.

2 50 kJ taken in (absorbed).

3 Decrease / lower it.

5.3

1 It will have a different number of molecules of gases in the reactants compared with the products.

2 High pressure produces more (higher yield of) ammonia, but very high pressures are expensive.

3 Low temperatures produce more (higher yield of) ammonia, but the reaction is too slow at lower temperatures.

End of chapter questions

1 Two from:
Combustion (burning fuels, burning metals) / respiration / neutralisation.

2 Endothermic reaction(s) happen when they mix with (dissolve in) the water *or* they react with water and take in heat / reactions happen that take in heat.

3 Lower / decrease the temperature.

4 95 kJ are taken in / absorbed.

5 No effect (because there are equal numbers of molecules of gases in the reactants and products).

6 Low temperature and high pressure.

Chapter 6

Pre Test

1 They are broken down (split up) into elements.

2 Non-metallic elements.

3 At the negative electrode.

4 A solution of an ionic compound in water in which the metal that formed the compound is more reactive than hydrogen.

5 Hydrogen, chlorine and sodium hydroxide.

6 The products have many important uses.

7 Copper atoms lose electrons / are oxidised.

8 They form a sludge (from which precious metals are extracted).

Check yourself

6.1
1 A molten ionic compound or a solution containing ions.
2 They gain electrons / are reduced.
3 Zinc and chlorine.

6.2
1 They are discharged / gain electrons / are reduced / form copper atoms.
2 They are discharged / lose electrons / are oxidised / form chlorine atoms / form chlorine molecules.
3 All metals form positive ions (so they are attracted to the negative electrode where they gain electrons).

6.3
1 Sodium / Na^+, hydrogen / H^+, chloride / Cl^-, hydroxide / OH^-.
2 Hydrogen ions and chloride ions are discharged, leaving sodium ions and hydroxide ions in the solution.
3 Hydrogen: making margarine, making hydrochloric acid.
Chlorine: sterilising water, making bleach or disinfectants or plastics.
Sodium hydroxide: making paper or soap or bleach, to control pH.

6.4
1 Impurities decrease conductivity / increase resistance of copper.
2 Copper atoms lose electrons / are oxidised to form copper ions.
3 Copper ions from the solution gain electrons at the negative electrode to form copper atoms that are deposited on the electrode.

End of chapter questions

1 They move towards the electrodes / positive ions are attracted to negative electrode and negative ions are attracted to the positive electrode.
2 Chlorine.
3 Magnesium ions gain electrons / are reduced to form magnesium atoms.
4 Bromine (Br_2).
5 Water (is present and) produces hydrogen ions that are discharged/reduced (rather than sodium ions).
6 Chlorine and sodium hydroxide.
7 They are oxidised / form copper ions (Cu^{2+}) that go into the solution.
8 It (is negatively charged so) attracts copper ions and they gain electrons / are reduced to form copper atoms that are deposited on the electrode.
9 $Al^{3+} + 3e^- \rightarrow Al$
10 $2Cl^- \rightarrow Cl_2 + 2e^-$
or
$2Cl^- - 2e^- \rightarrow Cl_2$

Chapter 7

Pre Test

1 (a) hydrogen ions / H^+(aq)
(b) hydroxide ions / OH^-(aq)
2 It is very alkaline / it contains a large amount of hydroxide ions / has a high concentration of hydroxide ions.
3 Salt and hydrogen.
4 Magnesium oxide or magnesium hydroxide.
5 Potassium sulfate and water.
6 By mixing solutions containing lead ions and chloride ions / by mixing solutions of a soluble lead salt (e.g. lead nitrate) with a soluble chloride (e.g. sodium chloride).

Check yourself

7.1
1 Hydrogen ions.
2 They produce hydroxide ions in the solution that make it alkaline.
3 Universal indicator or wide range indicator (*not* just indicator).

7.2
1 A salt and water.
2 So that all of the acid is used.
3 Zinc chloride.
4 Copper oxide and nitric acid.

7.3
1 An indicator.
2 Ammonium chloride.
3 Copper sulfate + sodium carbonate.

End of chapter questions

1 It is very acidic / it has a large amount (or high concentration) of hydrogen ions.
2 A base that dissolves in water and forms hydroxide ions in the solution.
3 Magnesium sulfate and hydrogen.
4 Copper chloride and water.
5 There is no visible change when acids react with alkalis.
6 Sodium nitrate and water.
7 H^+(aq) + OH^-(aq) $\rightarrow$ H_2O(l) (state symbols not essential)
8 Any named soluble sulfate, e.g. sodium sulfate (but *not* lead sulfate).

EXAMINATION-STYLE QUESTIONS

1 (a) One mark each for:
Suitable axes, correctly labelled.
All points correctly plotted (within half small square).
Smooth line through points.
Omitting 5 minutes / 65 cm^3. (4 marks)
(b) (i) Slope/gradient was steepest at start. (1 mark)
(ii) Two from:
Concentration of reactants highest at start / concentration of reactants decreases with time.
More particles of reactants at the start / reactants get used up over time.
Greater frequency of collisions at start / more collisions per second at start. (2 marks)
(c) Line with steeper slope initially, (1 mark)
levelling off at same volume as other line. (1 mark)

2 (a) So the ions can move / so the ions are free to move. (1 mark)
(b) Two from:
(Positive) sodium ions are attracted to the (negative) electrode.
Sodium ions gain electrons.
Sodium ions are reduced, form sodium atoms. (2 marks)
(c) Two from:
Chloride ions are attracted to the positive electrode.
Chloride ions lose electrons.
Chloride ions are reduced, form chlorine atoms / form chlorine molecules. (2 marks)
(d) So it does not react with chlorine. (1 mark)
(e) (i) Chlorine, sodium hydroxide. (2 marks)
(ii) Two from:
Water produces H^+ ions / both Na^+ and H^+ ions present.
Sodium is more reactive than hydrogen.
Hydrogen ions discharged/reduced in preference to sodium ions. (2 marks)

3 (a) (i) One mark each for:
Suitable axes, correctly labelled.
All points correctly plotted (within half small square).
Smooth lines through points (no daylight showing at points).
Temperatures correctly labelled. (4 marks)
(ii) 70 atmospheres (+/- 2) at 350°C (1 mark)
(iii) Line drawn between the two lines but closer to 500°C (1 mark)
(b) (i) Reversible reaction. (1 mark)
(ii) Yield better at low temperatures.
Reaction too slow at low temperatures / catalyst does not work at low temperatures. (2 marks)
(iii) High pressure gives high yield.
Very high pressures are too expensive / too dangerous / use too much energy / need very strong equipment. (2 marks)

Chapter 1

Pre Test

1 In order of atomic weights and by chemical properties.
2 Mendeleev.
3 They have the same number of electrons in the highest occupied energy level (outer shell).
4 They have different occupied energy levels (shells).
5 It increases / the elements become more reactive.
6 For elements lower in the group, the electron in the highest energy level (outer shell) is less strongly attracted by the nucleus / the electron in the highest energy level is further from the nucleus / the electron in the highest energy level is more shielded *and* so it is more easily lost / positive ions form more easily.
7 Green or coloured / gas / low melting and boiling point / made of (diatomic) molecules (Cl_2) / reacts with metals / forms ions with one negative charge (Cl^-) / forms covalent compounds (with non-metals) / has 7 electrons in its outer shell.
8 Decreases down the group.
9 They are metals / strong / hard / unreactive (with air and water) / have high melting points / can be alloyed / can be bent or beaten or forced into shape.
10 A lower energy level (inner shell) can hold ten more electrons (after Group 2) *or* the third energy level / shell can hold a total of 18 electrons.

Check yourself

1.1

1 Elements with similar properties occur at every eighth element when arranged in order of increasing atomic weights.
2 The seven groups contained elements with different properties / the pattern did not work.
3 He left gaps for elements that had not been discovered.
4 Some of the missing elements were discovered and shown to have the properties predicted by Mendeleev.

1.2

1 They have the same number of electrons in their highest occupied energy level (outer shell).
2 Caesium.
3 Chlorine.
4 Metals react by losing electrons and the electron in the highest energy level (outer shell) is more easily lost from larger atoms / atoms with more electrons / atoms with more occupied energy levels (shells) / atoms of elements lower in the group.

1.3

1 They produce an alkali / alkaline solution when they react with water *or* their hydroxides are (strong) alkalis.
2 They float (because they are less dense than water and other metals).
3 Lithium bromide, LiBr.
4 Potassium has more electrons or more occupied energy levels (shells) than sodium, so the electron in its highest occupied energy level (outer shell) is less strongly attracted by the nucleus and so is more easily lost.

1.4

1 They have small (diatomic) molecules and there are only weak (intermolecular) forces between the molecules.
2 Their atoms need to gain one electron and they can do this by forming an ion or by sharing electrons with another atom.
3 It has the smallest atom with the smallest number of electrons and the smallest number of occupied energy levels (shells), so it has the strongest attraction for (outer) electrons.
4 Add chlorine to a solution (in water) of a bromide salt (e.g. potassium bromide) and bromine will be produced (chlorine will displace bromine).

1.5

1 Between Groups 2 and 3.
2 Two from: all metals / all form positive ions / all have only a small number of electrons in their highest occupied energy level (outer shell).
3 Three from: less reactive / more dense / form more than one positive ion / coloured compounds / catalysts / unfilled lower energy level (inner shell).
4 An un-filled lower energy level (inner shell).

End of chapter questions

1 The elements had similar properties every eighth element when arranged in order of atomic weights or because of his law of octaves.
2 The elements fitted better into groups with similar properties because he left gaps for undiscovered elements.
3 Protons, electrons and the arrangement of electrons in atoms.
4 The atoms get larger because the number of electrons increases, so there are more occupied energy levels (shells), and the outer electrons are less strongly attracted by the nucleus.
5 (a) sodium + water → sodium hydroxide + hydrogen
  (b) $2Na + 2H_2O \rightarrow 2NaOH + H_2$
6 It has the smallest atom in the group, so the electron in its highest energy level (outer shell) is closer to the nucleus and more strongly attracted so it is less easily lost.
7 They are made of small (diatomic) molecules with only weak forces of attraction between the molecules.
8 (a) chlorine + potassium bromide → bromine + potassium chloride
  (b) $Cl_2 + 2KBr \rightarrow Br_2 + 2KCl$
9 It has two different ions, Cu^+ and Cu^{2+}, so it is a transition metal.
10 Two from: high melting and boiling points; hard; high density; strong; can be forced into shape; can be made into alloys.
Two from: unreactive or reacts slowly with oxygen (air) and/or water; forms positive ions; forms more than one positive ion; coloured compounds; catalyst.

Chapter 2

Pre Test

1 Acids that ionise completely in aqueous solution.
2 It has a lower concentration of hydrogen ions, $H^+(aq)$ / it has a higher pH [not just fewer hydrogen ions – must be concentration].
3 A method used to find volumes of solutions that react exactly / completely. (It uses a pipette and a burette to find reacting volumes of solutions.)
4 To show the end point / when the reaction is completed (because the solutions are colourless and there is no obvious change).
5 Concentrations of solutions (in moles per dm^3 or grams per dm^3) (and amounts of substances in solutions).
6 By multiplying by its relative mass / mass of one mole.

Check yourself

2.1

1 A hydrated proton or a hydrated hydrogen ion.
2 A strong acid (with a low concentration).
3 Two alkali metal hydroxides, e.g. sodium hydroxide, potassium hydroxide.
4 Test the pH of solutions with the same concentration – the weaker one has a lower pH.

2.2

1 A pipette.
2 To improve the reliability of the results.
3 So that the end point is sharp.
4 Phenolphthalein.

2.3

1 The volumes of the solutions that react together and the concentration of the sodium hydroxide solution.
2 (a) 2 moles KOH react with 1 mole H_2SO_4, so 0.1 mol per dm^3.
  (b) 5.6 g per dm^3.

End of chapter questions

1 It ionises completely in aqueous solution.
2 Either: compare its pH with a solution of a strong acid with the same concentration – its pH will be higher than the strong acid; or: compare its reaction with a metal with a strong acid of the same concentration – it will react slower than the strong acid.
3 Pipette and burette; the burette has a tap.
4 They are both weak and so will not give a precise or sharp end point.
5 That the solution of sodium hydroxide was twice the concentration of the acid or the acid was half the concentration of the alkali (because 1 mole of sodium hydroxide reacts with 1 mole of hydrochloric acid).
6 $20.0 \, cm^3$ (4.0 g per dm^3 is 0.10 mol per dm^3 and 1 mole nitric acid reacts with 1 mole sodium hydroxide).

Chapter 3

Pre Test

1 Only water evaporates (leaving behind dissolved salts) to water vapour that condenses to form rain.
2 A solution that contains the maximum amount of solute dissolved at a given temperature.
3 Line graphs of solubility against temperature.
4 Solubility decreases as temperature increases.
5 Dissolved calcium and magnesium salts or ions.
6 You need to use more soap, it causes scum, it produces scale that causes blockages and reduces efficiency of water heating systems.
7 By precipitating out calcium and magnesium ions from solution as carbonates.
8 Replacing metal ions in solution with other metal ions (that do not cause hardness).

9 Water that contains no harmful substances.

10 By sedimentation and filtering to remove solids and by killing microorganisms.

Check yourself

3.1

1 Evaporation and condensation.

2 Most ionic compounds, most gases, some molecular substances.

3 Grams of solute per 100 grams of solvent.

4 Solute (solid) (crystals) separates out or crystallises.

3.2

1 It increases with increasing temperature.

2 To show how solubility changes with temperature, to find solubilities at given temperatures, to find how much solute separates when the temperature of a saturated solution is changed.

3 Less oxygen dissolves at high temperatures.

4 Water that has carbon dioxide dissolved in it at higher than normal pressure.

3.3

1 Dissolved calcium and magnesium salts (compounds) or ions.

2 Salts or compounds or ions dissolve from rocks in contact with water.

3 It is good for healthy bones, teeth and heart.

4 Solid insoluble deposit produced when hard water is heated.

3.4

1 Water that does not contain any ions that cause hardness (calcium and magnesium ions or salts or compounds).

2 Adding washing soda (sodium carbonate) and ion exchange.

3 All metal carbonates except those of metals in Group 1.

4 By washing it through with a concentrated solution of sodium chloride.

3.5

1 Harmful substances.

2 To kill bacteria or microorganisms.

3 Charcoal, ion-exchange resin and silver or a substance to prevent growth of bacteria.

4 Removal of all ions from water (except for H^+ and OH^- ions).

End of chapter questions

1 Any two gases from: carbon dioxide, oxygen, nitrogen, argon, noble gases, sulfur dioxide, oxides of nitrogen.

2 The maximum amount of solute that will dissolve at the temperature of the solution.

3 Find solubility at 60°C, find solubility at 20°C, subtract the two solubilities.

4 By adding carbon dioxide to water under higher than normal pressure.

5 Calcium and magnesium ions (or salts or compounds) react with soap to form an insoluble solid (precipitate).

6 Causes blockages in pipes and reduces efficiency of water heating systems.

7 The metal ions are removed from solution so they cannot react with soap.

8 The ions that cause hardness are removed from the solution and replaced with ions that do not react with soap.

9 Use a suitable source, sedimentation, filtration, add chlorine or kill microorganisms.

10 Distillation or de-ionising.

EXAMINATION-STYLE QUESTIONS

1 (a) Ar and K or Te and I. (1 mark)
  (b) They are the same. (1 mark)
  (c) The number of electrons in the highest occupied level or outer shell is the same as the group number.
  (d) Relative atomic masses depend on the number of isotopes and the number of neutrons in the atom. (1 mark)

Chemical properties / group number depend on the number of protons / electrons. (1 mark)
  (e) Protons / electrons had not been discovered. (1 mark)

2 (a) Gases become less soluble as temperature increases. (1 mark)
    So they come out of solution or form bubbles. (1 mark)
  (b) It contains no calcium or magnesium ions, (1 mark)
    which react with soap. (1 mark)
  (c) Carbon dioxide and sulfur dioxide. (1 mark)
  (d) Two from: makes it hard; increases the pH or neutralises the acids; forms a precipitate or makes it cloudy (by reacting with carbon dioxide). (2 marks)
  (e) (i) Filtration or sedimentation. (1 mark)
    (ii) Chlorination or boiling. (1 mark)
      Kills bacteria / microorganisms. (1 mark)

3 (a) It increases as temperature increases. (1 mark)
  (b) Any value in the range 120–3. (1 mark)
  (c) Any value in the range 90–3 (dependent on (b)). (2 marks)

4 (a) It is a (weak) acid. (1 mark)
    So it affects the results / titration or reacts with sodium hydroxide. (1 mark)
  (b) (i) Decrease the volume (by boiling off water as steam). (1 mark)
    (ii) Systematic error: higher than they should be. (1 mark)
  (c) (i) Titrations 2 and 3 or 22.5 and 22.4. (1 mark)
    (ii) Closest together / most concordant / most accurate. (1 mark)
  (d) (i) Phenolphthalein (1 mark)
    (ii) Citric acid is a weak acid. (1 mark)
  (e) (i) 0.0045 moles. (2 marks)
    (ii) 0.0015 moles. (1 mark)
    (iii) 0.060 mol per dm³. (1 mark)
    (iv) 11.52 g per dm³. (1 mark)

Chapter 4

Pre Test

1 Simple calorimetry – burn known mass of substance and measure the temperature rise produced for a known volume of water.

2 Variables are difficult to control, heat losses difficult to avoid.

3 Bonds in reactants are broken and bonds in products are formed.

4 Energy axis, level for reactants, level for products, energy change of reaction, energy to break bonds, energy of making bonds.

5 The energy needed to break particular chemical bonds.

6 To calculate the energy needed to break or make bonds and to calculate the energy change of reactions.

Check yourself

4.1

1 A calorimeter.

2 The unit is kJ or J.

3 Carefully controlled conditions are needed or variables are difficult to control.

4 Mass of substances burnt, mass of water used, temperature change.

4.2

1 Bonds in the reactants must be broken.

2 The amount of energy needed to break the bonds of the reactants must be less than the energy released when bonds in the products are formed.

3 By the difference in the levels of reactants and products.

4 Measure volumes of reactants, measure

temperatures of reactants, mix in plastic cup, measure temperature of mixture, use results to calculate the energy change (4.2 J increase temperature of 1 cm³ of solution by 1°C).

4.3

1 The units are kJ per mole (kJ/mol).

2 The number of moles in balanced equation and the number of each type of bond in the quantities of the reactants shown in the balanced equation.

3 $\Delta H = -818$ kJ/mol (2648 – 3466).

End of chapter questions

1 A simple calorimeter, i.e. a metal can or glass beaker, a spirit burner for the fuel, also a thermometer, a balance and a measuring cylinder.

2 Much of the heat is lost and it is too difficult to control all the variables.

3 Energy for bond breaking is less than energy of bond making.

4 Diagram should have: energy axis on left-hand side increasing upwards, reactants on level below products, line or arrow upwards from reactants labelled 'bond breaking', line or arrow downward to products labelled 'bond making', line or arrow between reactants level and products level labelled 'energy change of reaction'.

5 Measure volume and temperature of water, measure mass of calcium oxide, measure final temperature after adding calcium oxide to water and mixing well.

6 −103 kJ/mol.

Chapter 5

Pre Test

1 Flame test and adding sodium hydroxide solution.

2 Transition metal ions or Cu^{2+}, Fe^{2+} and Fe^{3+}.

3 Add dilute hydrochloric acid, test gas given off with lime water, turns milky if carbonate ions present.

4 Add dilute nitric acid and silver nitrate solution. A precipitate shows halide ions are present.

5 They burn or char.

6 By burning a known mass of the compound and using the masses of the products.

7 Atomic absorption spectroscopy and mass spectrometry.

8 Advances in electronics and computing.

9 Chromatography.

10 Two from: infra-red spectroscopy, nuclear magnetic resonance (NMR) spectroscopy, mass spectrometry.

Check yourself

5.1

1 Lithium Li^+, sodium Na^+, potassium K^+, calcium Ca^{2+}, barium Ba^{2+}.

2 Magnesium Mg^{2+}, calcium Ca^{2+}, aluminium Al^{3+}.

3 Copper(II) Cu^{2+}, iron(II) Fe^{2+}, iron(III) Fe^{3+}.

4 Add sodium hydroxide solution, warm, test gas with damp red litmus paper – turns blue if ammonium ion present.

5 $Cu^{2+}(aq) + 2OH^-(aq) \rightarrow Cu(OH)_2(s)$
  or $Fe^{2+}(aq) + 2OH^-(aq) \rightarrow Fe(OH)_2(s)$
  or $Fe^{3+}(aq) + 3OH^-(aq) \rightarrow Fe(OH)_3(s)$

5.2

1 Chloride, bromide, iodide and sulfate.

2 To show that it is carbon dioxide (and not other gases).

3 Ammonium ions would produce ammonia gas with sodium hydroxide and turn the litmus paper blue.

4 It decomposes to give an oxide that is yellow when hot but white when cold.

5.3

1 They are organic and contain carbon (and do not burn).
2 Mass of substance burnt, mass of each product (also relative atomic masses).
3 It reacts with carbon–carbon double bonds and is decolourised.
4 Pipette, burette, conical flask, iodine solution of known concentration, solution of unsaturated oil of known concentration or known mass of oil.

5.4

1 They are more rapid, sensitive and accurate and use only small samples.
2 Electronics in developing the instruments themselves and sample handling equipment, computers for control of instruments and for processing data (also to display or present results).
3 Detecting (and measuring) very small amounts of elements (mainly metals).
4 To identify and measure the amounts of elements in substances, to identify isotopes and their ratios in substances, to find relative atomic masses.

5.5

1 To separate the compounds in a mixture.
2 (Different types of) radiation (or several names, e.g. IR, UV, visible light, radio waves).
3 Three from: infra-red, ultraviolet (and/or visible light), nuclear magnetic resonance, atomic absorption spectroscopy.
4 Infra-red, nuclear magnetic resonance, mass spectrometry.

End of chapter questions

1 Calcium ions, Ca^{2+}.
2 Ammonium ions, NH_4^+.
3 Copper(II) carbonate.
4 Iodide, I^-.
5 Add bromine water and shake, changes from orange/brown to colourless.
6 CH_3.
7 To control instruments, to automate sample handling, to process data, (to compare data, to display or present results).
8 Uses very small samples, can detect very small amounts, can detect many elements (metals).

9 To separate the compounds in a mixture before they are identified or analysed further.
10 Infra-red (IR) spectroscopy.

EXAMINATION-STYLE QUESTIONS

1 (a) Chocolate (bar). (1 mark)
  (b) Low fat crisps. (1 mark)
  (c) 5031 kJ (2 marks)
    (Working: 1 mark for $(956 \times 2.5) + (1174 \times 1.5) + (176 \times 5)$)
    3655 kJ (2 marks)
    (Working: 1 mark for $(2095 \times 1.2) + (2116 \times 0.25) + (153 \times 4)$)
  (d) 50 g (1 mark)
    42.9 g (1 mark)
  (e) Either: less energy (1 mark)
    so less likely to gain weight/become obese (1 mark)
    or: less fat (1 mark)
    so less risk of heart disease (1 mark)
  (f) Either: may need to gain weight or may need extra energy
    or: the percentage of fat / proportion of fat is higher in the friend's suggestion
    or: other valid suggestion, e.g. may contain less carbohydrate or other nutrients/vitamins. (1 mark)

2 Several possible schemes so give one mark for each test, one mark for each result and correct conclusion up to max. 8 marks:
  ● Add sodium hydroxide solution to all four solutions (1): two solutions give white precipitates (1) warm the solutions with no precipitate (1): one gives off gas that turns damp blue litmus red = ammonium sulfate (1), the other solution is potassium sulfate (1). Test fresh solutions of the two that gave precipitates in flame test (1): red colour is calcium chloride (1), the other is magnesium chloride (1).
  ● Add dilute hydrochloric acid and barium chloride solution to all four solutions (1): two give white precipitates = sulfates (1) the other two are chlorides (1). Test fresh solutions of sulfates with flame test (1) lilac colour = potassium sulfate (1), the other is ammonium sulfate (1). Test

chlorides with flame test (1) red = calcium chloride (1). (Alternative for sulfates: add sodium hydroxide solution and warm (1) one turns damp red litmus blue = ammonium sulfate (1), the other is potassium sulfate (1).
  ● Add dilute nitric acid and silver nitrate solution to all four solutions (1): two that give precipitates are chlorides (1) the others are sulfates. Test chlorides and sulfates as above. (8 marks)

3 (a) Yellow. (1 mark)
  (b) One mark for metal ion, one for correct colour:
    lithium – red; potassium – lilac; calcium – red; barium – green. (2 marks)
  (c) Each element produces a unique pattern (of radiation). (1 mark)
  (d) The amount of radiation depends on the amount of the element in the solution. (1 mark)
  (e) Any two from: (very) small sample; accurate; rapid. (2 marks)

4 (a) One mark each for a diagram with some labels showing: burner or sample burning; under a can or beaker; containing water and a thermometer. (3 marks)
    One mark each for: mass of sample burnt; mass or volume of water; temperature change of water. (3 marks)
  (b) (i) 4728 kJ (2 marks)
      (Working: $5 \times 413 + 358 + 464 + 347 + 3 \times 498$) (1 mark)
    (ii) 6004 kJ (2 marks)
      (Working: $4 \times 805 + 6 \times 464$) (1 mark)
    (iii) 1276 kJ (1 mark)

5 (a) Unsaturated or contains a carbon–carbon double bond. (1 mark)
  (b) CH_2 (3 marks)
    (Working: $CO_2 = 44$ and $H_2O = 18$ (1 mark); 1.10/44 and 0.45/18 or ratio 0.025 C and 0.05 H. (1 mark)
  (c) C_4H_8 (1 mark)

ASSESSMENT and
QUALIFICATIONS
ALLIANCE

1. **Reactivity Series of Metals**

Potassium	most reactive
Sodium	
Calcium	
Magnesium	
Aluminium	
Carbon	
Zinc	
Iron	
Tin	
Lead	
Hydrogen	
Copper	
Silver	
Gold	
Platinum	least reactive

(elements in italics, though non-metals, have been included for comparison)

2   **Formulae of Some Common Ions**

Positive ions

Name	Formula
Hydrogen	H^+
Sodium	Na^+
Silver	Ag^+
Potassium	K^+
Lithium	Li^+
Ammonium	NH_4^+
Barium	Ba^{2+}
Calcium	Ca^{2+}
Copper(II)	Cu^{2+}
Magnesium	Mg^{2+}
Zinc	Zn^{2+}
Lead	Pb^{2+}
Iron(II)	Fe^{2+}
Iron(III)	Fe^{3+}
Aluminium	Al^{3+}

Negative irons

Name	Formula
Chloride	Cl^-
Bromide	Br^-
Fluoride	F^-
Iodide	I^-
Hydroxide	OH^-
Nitrate	NO_3^-
Oxide	O^{2-}
Sulfide	S^{2-}
Sulfate	SO_4^{2-}
Carbonate	CO_3^{2-}

3. The Periodic Table of Elements

Key

relative atomic mass
atomic symbol
name
atomic (proton) number

Example:

1
H
hydrogen
1

1	2											3	4	5	6	7	0
						1 **H** hydrogen 1											4 **He** helium 2
7 **Li** lithium 3	9 **Be** beryllium 4											11 **B** boron 5	12 **C** carbon 6	14 **N** nitrogen 7	16 **O** oxygen 8	19 **F** fluorine 9	20 **Ne** neon 10
23 **Na** sodium 11	24 **Mg** magnesium 12											27 **Al** aluminium 13	28 **Si** silicon 14	31 **P** phosphorus 15	32 **S** sulfur 16	35.5 **Cl** chlorine 17	40 **Ar** argon 18
39 **K** potassium 19	40 **Ca** calcium 20	45 **Sc** scandium 21	48 **Ti** titanium 22	51 **V** vanadium 23	52 **Cr** chromium 24	55 **Mn** manganese 25	56 **Fe** iron 26	59 **Co** cobalt 27	59 **Ni** nickel 28	63.5 **Cu** copper 29	65 **Zn** zinc 30	70 **Ga** gallium 31	73 **Ge** germanium 32	75 **As** arsenic 33	79 **Se** selenium 34	80 **Br** bromine 35	84 **Kr** krypton 36
85 **Rb** rubidium 37	88 **Sr** strontium 38	89 **Y** yttrium 39	91 **Zr** zirconium 40	93 **Nb** niobium 41	96 **Mo** molybdenum 42	[98] **Tc** technetium 43	101 **Ru** ruthenium 44	103 **Rh** rhodium 45	106 **Pd** palladium 46	108 **Ag** silver 47	112 **Cd** cadmium 48	115 **In** indium 49	119 **Sn** tin 50	122 **Sb** antimony 51	128 **Te** tellurium 52	127 **I** iodine 53	131 **Xe** xenon 54
133 **Cs** caesium 55	137 **Ba** barium 56	139 **La*** lanthanum 57	178 **Hf** hafnium 72	181 **Ta** tantalum 73	184 **W** tungsten 74	186 **Re** rhenium 75	190 **Os** osmium 76	192 **Ir** iridium 77	195 **Pt** platinum 78	197 **Au** gold 79	201 **Hg** mercury 80	204 **Tl** thallium 81	207 **Pb** lead 82	209 **Bi** bismuth 83	[209] **Po** polonium 84	[210] **At** astatine 85	[222] **Rn** radon 86
[223] **Fr** francium 87	[226] **Ra** radium 88	[227] **Ac*** actinium 89	[261] **Rf** rutherfordium 104	[262] **Db** dubnium 105	[266] **Sg** seaborgium 106	[264] **Bh** bohrium 107	[277] **Hs** hassium 108	[268] **Mt** meitnerium 109	[271] **Ds** darmstadtium 110	[272] **Rg** roentgenium 111							

Elements with atomic numbers 112–116 have been reported but not fully authenticated

* The Lanthanides (atomic numbers 58 –71) and the Actinides (atomic numbers 90–103) have been omitted.

Cu and **Cl** have not been rounded to the nearest whole number.

114